Anonymous

The Standard American Poultry Book

Containing All the Different Varieties of Fowls

Anonymous

The Standard American Poultry Book
Containing All the Different Varieties of Fowls

ISBN/EAN: 9783337812744

Printed in Europe, USA, Canada, Australia, Japan

Cover: Foto ©berggeist007 / pixelio.de

More available books at **www.hansebooks.com**

STANDARD Price, 25c.
American
Poultry Book.

aining descriptions of all the Different Varieties of Fowls, with complete instructions for raising all kinds of Poultry, Curing Diseases, Artificial Incubation, etc., etc.

W. L. ALLISON, PUBLISHER,
216-218 WILLIAM ST.—18-20 ROSE ST.,
NEW YORK.

THE STANDARD
American Poultry Book,

CONTAINING ALL THE DIFFERENT

Varieties of Fowls,

Their Points of Beauty, their Merits as Layers or Sitters;

WITH COMPLETE INSTRUCTIONS

ON RAISING ALL KINDS OF POULTRY—THE BEST SOILS ON WHICH TO KEEP THEM
—HOW TO FEED—MANAGEMENT OF LAYERS AND SITTERS—DICTIONARY
OF POULTRY TERMS—INCUBATION—MANAGEMENT OF THE MOTHER—
HOW TO REAR THE CHICKS — IMPROVED MANNER OF

CURING ALL DISEASES.

Together with minute Instructions on

ARTIFICIAL INCUBATION,

The Best Incubators—How to work them, etc.,

Copyrighted 1899 by THE CHISWICK PUBLISHING CO.

NEW YORK:

W. L. ALLISON, PUBLISHER.

Entered at the Post Office, New York, N. Y., as second-class matter,
January, 1899.

PLYMOUTH ROCKS.

DOMESTIC POULTRY.

VARIETIES; THEIR CHOICE AND MANAGEMENT.

In the choice of fowls, no inconsiderable amount of knowledge of the characters of the different varieties is necessary to insure success to the breeder. From my own experience, and that of the most eminent poultry-keepers, I have attempted to jot down such information as may be found useful in the selection and management of these really useful and elegant birds. I shall first introduce to the reader's notice the largest and one of the most important breeds in our country.

THE BRAHMA FOWLS

Are divided into two classes, the light and dark; as a rule the dark are preferable, although either are good enough for any farm yard. They are now almost universally cultivated throughout America, and a most valuable variety—so hardy, so beautiful, and so excellent in all the relations of poultry life.

The hens are the best of mothers, and lay fine large eggs during the winter. Even when the ground is covered with snow, they lay regularly, and in fact at all times when not employed in sitting or renewing their plumage. The pullets attain full size at an early age, and are in their prime when eight months old.

Brahmas are doubtless the largest of all the varieties of domestic fowls; some have been known to weigh seventeen pounds, which exceeds the weight of any other breed.

LIGHT BRAHMA, COCK AND HEN.

The dark Brahmas have steadily progressed in favor since their first introduction; their gigantic size, great weight, hardihood and prolificacy, and the ease with which they can be kept in confined ranges, all tend to render them much esteemed. To sum up their merits, as good, useful, hardy fowls, they are unsurpassed. They are good layers of good sized eggs, good foragers and good sitters; as mothers they cannot be excelled, no fowls being more careful not to step on their chickens, brooding them better, or searching more diligently for food. The chickens grow fast and are exceedingly hardy; old and young take good care of themselves, and often recover from ailments that would carry off any of a less hardy sort. They are very good for the table, putting on flesh readily; they are also small eaters.

DARK BRAHMA HEN.

DESCRIPTION OF THE DARK BRAHMA.

The head of the dark Brahma cock should have a pea comb, that is a triple comb; this should be small, low in front, and firmly set in the head without falling over on either side, distinctly divided, so as to have the appearance of three small combs joined together in the lower part and back, the largest being in the middle, and each part slightly and evenly serrated.

The upper part of the body is silvery white striped with black; the breast, under part of body and thighs either pure black or slightly mottled with white. The feathers that cover the bases of the quill-feathers of the wings are of a lustrous green black, and form a broad well-marked bar across the wings. The flight-feathers are white on the outer and black on the inner webs. The secondary quills have a broad, dark, green black spot at the end of each feather. The tail is black. The shank should be of a yellow color, and well clothed with dark feathers slightly mottled with white.

The hens have a grey head; neck-hackle silvery white, striped with black. The comb is the same only of a smaller size. The remainder of the plumage should be dull white, and closely pencilled with dark steel grey so as almost to cover the ground color and reaching well up the front of the neck. The hen is not so upright in carriage as the cock, and it is much shorter in the legs.

LIGHT BRAHMAS.

LIGHT BRAHMA HEN.

In color, the light Brahmas are characterized by the general white color of the body, breast and thighs. The neck-hackle should be marked with a distinct black stripe down the centre of each feather; there is a tendency in the cock to come light or cloudy in the hackle—defects which very greatly detract from their beauty. The saddle-feathers in the cock are white or lightly striped with black, those of the hen being white. The first ten flight-feathers are black, but the secondary quills, which alone are visible when the wing is closed, are white on the outer web, consequently the dark color of the wing is not visible when folded. In the cock the tail is black, the tail-coverts being beautifully glossed with green, the lower ones being margined with

LIGHT BRAHMA COCK.

silver, as are the two highest tail-feathers in the hen. The shanks in this variety should be bright yellow, well closed with white feathers slightly mottled with black.

In conclusion I would state that I do not believe, all things considered, that there is any better market fowl than the Brahma; some other breeds are doubtlessly superior table fowls, but they are more tender and harder to rear.

Give your Brahmas large roomy quarters in winter, and if possible plenty of range for exercise; feed well, and they will give you winter-eggs, and those are the kind that bring money.

COCHINS.

BUFF COCHINS.

The Cochins were first introduced into this country under the name of Shanghaes; they originally come from Shanghae, and are to this day found in great numbers there. But the Shanghaes, as originally introduced and widely grown in this country, were gigantic muscular birds of great activity and wonderful powers of storing away food, which was absorbed into muscle and bone, but made comparatively little juicy flesh. The consequence was, they got a bad reputation, and the term was finally one of reproach; but upon the vast improvement which was made in them by careful breeding, the name of Cochins, as designated by cinnamon or white or buff or partridge, rapidly superseded the old term, and the despised but vastly improved Shanghae sailed under a new name, and are now raised as profitable birds all over the United States.

They are first-class layers, and in season when new-laid eggs are rare, and from their scarcity of so much increased value, this species often proves a source from whence we can obtain supplies. They also make capital mothers, and are quiet when sitting.

In many places where space is limited, the Cochins are found convenient guests; they can put up with worse accommodation, and require less space than almost any other race. I do not mean to say that they will thrive the better for confinement, neither that fowls in general will pine and die if kept in a narrow range; all fowls are better for having ample space; but in cases where their liberties are necessarily abridged and more careful tending is required to counterbalance want of field-room, the Cochin can bear captivity better than any other fowl.

The roosting-poles for Cochins and in fact all bulky fowls should be near the ground; they should be large in diameter in order that the claws may maintain a firm clutch and perfect equilibrium without inconvenience or effort.

VARIETIES OF THE COCHIN.

The Cochin fowl is a large heavy bird, very broad and clumsy looking. The tail is very short and nearly destitute of feathers, but the remainder of its body is abundantly covered. The legs are short, stout, and well feathered; the head should be small, with a single straight comb; the beak short and strong; the wattles small, and the ear-lobes red and fine as to texture. There are many varieties of the Cochin viz—BUFFS; this is the true type of the colored birds, and for utility, I think the best.

PARTRIDGE COCHINS.—Very heavy; full round plump forms and a majestic carriage.

WHITE COCHINS.—These should be pure white all over. In city yards amid the smoke and dust the White Cochin do not appear to advantage but in the country no variety looks more pleasing, as the beauty of their plumage depends on its clean and unsullied condition.

If well selected, properly taken care of, and well fed, they make a first-class table bird; they are hardy, do not require much space, and I should recommend them to any person who only wants to keep a few birds.

SPANISH FOWLS.

The Spanish fowls have long been known and highly esteemed in the United States for their great laying and non-sitting propensities.

All fowls are better for being hatched in a warm season, and the Spanish are no exception to this. Though of a sound constitution, no fowl is more injured by cold wet weather. Their

BLACK SPANISH FOWLS.

roosting-places therefore should face the south, and be well-protected from cold winds, especially as they are subject to long and protracted moultings. The cold affects their comb also, which is sometimes frost-bitten, with a liability to mortification.

The flesh of the Spanish Black fowl is juicy and of good flavor, but not equal to that of the "Dorking." The flesh of the White Spanish is not considered so fine in flavor, as that of the Black, yet it is not bad, especially if young.

As layers they are among the best, but are seldom inclined to sit; they generally produce two eggs consecutively and then miss a day.

As to healthiness, they are less liable to roup than lighter-colored birds; in fact, the Spanish fowl is less subject to disease than are most of the common black varieties.

In general they are rather quarrelsome, and are very averse to strange fowls, and if separated from each other even for two or three days, the hens will disagree seriously upon being reunited.

In case of a strange hen being tormented by her companions for any length of time, so that she is afraid to come and feed with them, or of the cock displaying his protracted dislike to her, it will be right to remove her, or she may be reduced to so low a condition as to render her unable to escape their persecutions, and avoid death from their violence.

The Spanish pullets commence laying when six or seven months old, and occasionally sooner, though some of them commence at a later period, according to feeding and treatment. But premature fertility is not to be wished for, as it will frequently happen that pullets which commence very early, seldom lay when fully grown so large an egg as those produce which do not lay before they are eight months old. Indeed the debilitating effects of either premature, or continual laying in ripe age, as respects the Spanish breed are now and then manifested by the loss of the body feathers in moulting, besides the usual falling off of the neck, and wing, and tail feathers; and when thus stripped, the poor birds look very miserable in bad weather.

In sitting Spanish eggs, nine of them are sufficient for hens of ordinary size, as they are much larger than the generality of fowls eggs.

It will be unwise, with any breed, to select the first dozen of a pullet's eggs for hatching; they being comparatively immature and small, it is not likely that large and strong chicks will be the issue. Besides, pullets occasionally do not enter into tender union with their male companions until they have laid five or six eggs.

The color of the Spanish chick, when first hatched, is a shining black, with a blotch of white sometimes on the breast, and a little white also around the bill and the eyes. They do not until nearly grown, get their full feathers, and therefore they should be hatched at a favorable season of the year, to be well feathered before it grows cold in the Fall.

Spanish hens seldom exhibit a disposition to undertake the task of incubation, and if it be attempted, they will in the generality of cases forsake the nest long before the chicks would be hatched. Sometimes, however, they will perseveringly perform the maternal duties; but it is against their general character. They are exceedingly long in the leg, consequently are subject to cramp; this partly accounts for their being so averse to such sedentary occupation. Since, therefore, they will not undertake the office of mothers, we must impose it upon some other class of fowl, that will not only accept the task, but will joyfully hatch and rear the young of even another species until they are able to take care of themselves. It is by this means the Spanish breed is still preserved and multiplied.

VARIETIES AND DESCRIPTION OF THE SPANISH FOWL.

A full grown black Spanish cock weighs seven pounds; the hen, about six pounds. The principal features, and those which form the most striking contrasts to those of other fowls, are, its complete suit of glossy black, large face, and ear-lobe of white; enlivened by comb and gills of excessive development. The peculiarities of these contrasts induce me to describe them in detail. The plumage is of a rich satin, black, reflecting their shades of bluish, greenish purple, when exposed to the sun's rays; the feathers of the breast, belly, and thighs, are black, of the most decided hue. The hens are of a similar feather, but less brilliant. The face and ear-lobes especially the latter, are of pearly whiteness; the face should extend above the eye, encircle it, and meet the comb; it still increases as the bird grows older, continuing to enlarge in size, especially with hens, which seldom have a really good show of face until two years of age, even beyond the time of their full growth; and the more face and ear-lobe, the more valuable either the cock or hen. The comb of the cock should be erect and serrated, almost extending to the nostrils, and of bright scarlet; it should be fine in texture, and exhibit no sign of excrescences. In hens this uprightness of comb cannot be obtained, owing to its abundant size and thinness of base. The wattles are long, pendulous, of high color, and well folded. The head is long, and there should be no topknot behind the comb, nor muff round the neck. The beak is long, and generally black, it should be slightly curved, and thick at the base. The eyes are very full, bright and of a rich chestnut color: they are somewhat prominent. The neck is rather long, but strong and thick towards the base, the neck hackle being a glossy black; the chest and body are broad and black, the former being particularly dark; the wings are of a moderate size, whilst the coverts are beautifully shaded, and of a bluish black. The thighs are neat but long, as also is the shank, which is of a leaden or dark blue color, and sometimes of a pale blue-white. The soles of the feet are of a dingy flesh-color; the tail is rather erect and well balanced, presenting if well plumed (as it should be) a very elegant green hued shade.

WHITE SPANISH.—These birds are not so hardy, but they inherit the usual qualities of the black; the general feathers, like the face being perfectly white.

THE ANCONA.—There is seldom much white about the face of this variety, and in many cases none; the ear-lobes is, however, of that color, though not so long and full as in the Black. They possess the general characteristics of the Spanish class, and are excellent layers. They are of a very unsettled color, spotted with white but far from regularly marked; they also present many ther shades and colors.

MINORCAS.—These are very similar to the last named variety, wanting the white face of the Black tribe; the shank is not so .ong as in the true Black. They are good layers, but bad sitters and mothers.

ANDALUSIAN.—When carefully selected, the chicks throw black and white and if those most resembling the originals are bred together, a neat grey bird may be obtained. They are good layers, and far better sitters and mothers than the Blacks, and have shorter shanks; whilst their principal peculiarity consists in a tail standing very erect, the feathers of which in many specimens nearly touch the hackle-feathers of the neck. They are a very hardy fowl, and possess a fair share of the Black's good qualities.

There are many other sub-varieties, or rather strains, that have crossed with the Spanish stock, but they neither deserve nor possess a distinct name.

The superiority of the Spanish generally, as egg producers, is so decided, that any cross from them meriting the character of everlasting layers, is worth encouragement. It is to be recollected that the Hamburgh or Dutch is not the only sort from which everlasting layers have sprung. Any hens which with warmth and good feeding will lay eggs continuously, and especially through the winter, are to be welcomed. And though the debilitating effects of continued laying must tell upon the constitution, yet where stock is not desired for a mere gratification of the eye, but kept on economical principles, it cannot be inexpedient to stimulate the prolific powers of hens to the utmost. If good layers which have not the presumption to compete for the prizes of birth or beauty, can by clever management, be induced to lay within two years the entire compliment of eggs which in the ordinary course of nature would not be yielded by them in less than three years, there is an actual saving gained of at least one-third of food, if these effete layers be then fattened and killed. No breed would be better if this plan is strictly applied, than that of the common Blacks of Spanish blood, or some of their sub-varieties.

THE DORKING FOWL.

Of distinct English breeds the Dorkings are the most celebrated. For those who wish to stock their poultry-yard with fowls of the most desirable shape and size, clothed in rich and variegated plumage, and not expecting perfection are willing to overlook one or two other points, the speckled Dorkings are the breed to be at once selected. The hens, in addition to their gay

DORKING FOWL.

GRAY ENGLISH DORKINGS.

colors, have a large vertically flat comb, which, when they are in high health, adds very much to their brilliant appearance. The cocks are magnificent; the most gorgeous hues are frequently lavished upon them, which their great size and peculiarly square-built form displays to great advantage. The breeder, and the farmer's wife, behold with delight their broad breast, the small proportion of offal, and the large quantity of profitable flesh.

The Cockerels may be brought to considerable weight, and the flavor and appearance of the meat are inferior to none. They are only fair layers, but at due and convenient intervals manifest the desire of sitting. Having short, compact legs, they are well formed for incubation. The Dorkings are not well suited for damp soils, by reason of the shortness of their legs. They are also distinguished for breadth of body, the somewhat partridge form, and also, in the poultry phrase, for being clean headed. Though they possess great similarity of form, there is much variety of color; but they are generally distinguished as white, grey or speckled, and also by the character of the comb—viz, as single and double, or rose combed; and classed accordingly at the poultry shows.

The fifth or supernumerary toe is the peculiar mark distinctive of the whole breed under consideration. Though the Creator has not designed anything without its appropriative purpose, this additional member must rather be deemed a distinctive than a useful one, just as the absence of a tail, or the color and size of a comb may distinguish an individual race of fowls. These over-furnished claws have been denounced as sources of danger and annoyance to young chicks when first issuing from the shell, rendering the mother's movements hazardous to them. I have never seen them do so, and even if they did how is the hen to be employed when the sitting fit comes on, for they are persevering sitters, and as neither worrying, nor whipping, nor fettering, nor physicking, or the cold shower bath, will subdue their natural instinct to set, they should be allowed to follow their instinct, and incubate in peace.

The Dorkings are a very heavy fowl when fat, as their frame work is not of that lengthy, incompact structure which it is so difficult to fill up with flesh and fat; they much sooner become tempting figures for trussing and skewering than other fowls. They have a great aptitude for fattening when rendered capons.

VARIETIES AND DESCRIPTION.

WHITE DORKINGS.—This variety seldom produces more than two broods a year, because they require more favorable seasons, and greater warmth than the colored.

The white is not so large as the colored, and, as a general rule, whiteness in animal physiology is indicative of constitutional delicacy. Their average weight is less than that of the colored, and like all white feathered poultry, the flesh has a tendency to yellowness.

The white cock and hen are perfectly white in the plumage, bills, and legs; both should have a double or rose-comb of bright red, though a single one is frequent, but this is considered a sign of degeneracy. The cock is very upright and spirited in his appearance, and his spurs are usually lower than those in other species. The fifth toe should be well defined. The hen has no individualities.

THE GREY OR SPECKLED DORKING COCK.—The head round, and furnished with double or single comb, of bright red; wattles, large and pendent; the ear-lobes almost white; hackles, a cream white, and the feathers of the hackles dark along the centre; the back, grey of different shades, interspersed with black; saddle feathers, same as hackles in color; wing feathers, white, mixed with black; the larger wing coverts, black; the lesser, brown and yellow, shaded with white; breast and thighs, black or dark brown; tail feathers, very dark, with a metallic lustre.

THE GREY OR COLORED HEN.—Face, lighter colored than that of the cocks; hackles, black and white; back, dark grey; saddle and wing, grey, tipped with black; tail, almost black. Five claws and white legs characterize both sexes.

POLAND FOWLS.

WHITE CRESTED BLACK POLISH COCK AND HEN.

The Polands are excellent layers of perfectly white and moderately-sized eggs, much pointed at the smaller end. They seem to be less inclined to sit than any other breed, and it is judicious to put their eggs under other nurses. The chicks of both sexes, which are hardly distinguishable for many weeks, are very ornamental. The male bird is first distinguished by the tail remaining depressed, awaiting the growth of the sickle feathers, whereas the female carries it uprightly from the first· also, the top-knot in the cockerels hangs more backward than in the pullets.

Their flesh is excellent, being white, tender and juicy.

During three or four years the cocks in particular increase in size, hardihood, and beauty, different in this from fowls generally, which advance much more rapidly to their highest points of perfection, but from which they fall away with corresponding rapidity.

The Polands are extremely tender, and so difficult to rear, that the eggs should not be set before the middle of May, as dampness is fatal to them while very young; but, if they live to be adults, no fowls are more hardy, or profitable as layers, or more delicious for the table.

Their demerits are few, and of no serious importance. They are not at all suited to dirty farmyards, becoming blind and miserable with dirt. They do not lay quite so early in the year as other tribes, and are not suited for the office of mothers and nurses, from their great disposition to lay; and when they do sit, they are rather unsteady and perverse. Now these objections may be dismissed, because there is nothing to prevent the substitution of hens of other tribes for hatching, and if the Polish hens and pullets themselves in the mean time lay eggs, there is no loss in an economical point of view.

We have good practical authority for stating that the critical period of their lives is from the second to the sixth month.

DESCRIPTION AND VARIETIES.

The crest of the cock is composed of straight feathers, something like those of a hackle or saddle; they grow from the centre of the crown and fall over outside, forming a circular crest. That of the hen is made up of feathers growing out and turning in at the extremities, till they form a large top-knot, which should in shape resemble a cauliflower. The comb of the cock is peculiar, inasmuch as it is very small, scarcely any on the top of the head, and having in front two small spirals or fleshy horns. The carriage is upright, and the breast more protuberant than in any other fowl, save the Sebright bantam. The body is very round and full, slightly tapering to the tail, which is carried erect, and which is ample, spreading towards the extremity in the hen, and having well defined sickle feathers in the cock. The legs should be lead color or black, and rather short than otherwise.

POLISH FOWLS.

The varieties among us are the Black; White; the Golden Spangled; and the Silver Spangled.

BLACK POLANDS.—Cock; body, neck, and tail, black, with metallic tints of green; crest, white, with a few black feathers at the base of the bill; comb, very small, consisting only of two or three spikes; large wattles, bright red; ear-lobe, white; the skull, instead of being flat as in other varieties, has a fleshy protuberance or round knob.

HEN; the same colors; wattles smaller than those of the cock; in other points the same.

WHITE POLANDS.—These should be pure white all over with the exception of the legs which are of a blue or slate color.

GOLDEN SPANGLED.—Cock; ground color, very bright ochre yellow, black spangles, which, in a particular light, have a beautiful greenish tint; crest, chestnut, with a few white feathers, black beard; comb and wattles small; hackle and saddle feathers, golden yellow; thigh, generally black, but some specimens have them spangled; sickle feathers, dark brown and very large, the smaller side ones lighter in the colors, and beautifully faced with black; legs, slate color.

HENS;—general colors the same; breast, neck, and back, spangled; tail and wing feathers, laced.

SILVER SPANGLED.—The only difference between this variety and the preceding one is in the ground, which is a beautiful silver white.

The Polands very often have crooked backs; when buying them the best mode for detecting the deformity is to lay the palm of the right hand flat on the bird's back, by which any irregularity of either hip, or a curve in the back bone from the hips to the tail will be detected.

THE SULTAN FOWL.

The Sultans, or Feather-footed White Polish, are a very elegant and pleasing variety, and were imported from Constantinople. They partake of the character of the Polish in their chief characteristics, in compactness of form and good laying qualities.

In general habits they are brisk and happy tempered. They are very good layers of large white eggs, but are non-sitters and small eaters.

As adults they are very hardy, with the exception of the tendency to cold, to which all crested birds are subject when exposed; but the chickens, from their rapid and early feathering, are difficult to rear, evidently suffering severely from the extra strain on their young constitutions.

DESCRIPTION.

In form they are very plump, full crested, short-legged and compact; the plumage pure and unsullied white throughout and very abundant; their tails are ample, and carried erect; their thighs are short, and furnished with feathers which project beyond the joint, or vulture hocked. Their legs are short, white, and profusely feathered to the feet, which are five toed. The comb consists of two small spikes situated at the base of a full-sized globular Polish crest; the wattles are small and red, wrinkled, both sexes being amply bearded. No fowls are more abundantly decorated—full tail of sickle-feathers, abundant furnishing, boots, vulture-hocks, beards, whiskers, and full round Polish crests, formed of closely-set, silky, arched feathers, not concealing the eyes, but leaving them unobscured.

The legs, as old age approaches, are apt to get red, swollen and inflamed, perhaps from the spur growing in a curved form and producing irritation.

All the varieties of the Polish if kept in a damp situation are liable to a cold, apt to degenerate into roup, and if they are too closely bred, liable to tuberculous diseases and deformity of the spine, causing humpback, they are also very subject to vermin unless supplied with a sand bath; vermin, however, may be readily destroyed by dusting flour of sulpher under the feathers with a common flour-dredger.

THE MALAY FOWL.

The Malay is a large heavy fowl, with close fitting plumage; it stands very high, and has an upright carriage; height is considered a great point in this breed; the head is small for the size of the bird, with considerable fulness over the eye, which should be pearl, and the hawk bill should be quite free from stain. Like the game fowl, the Malays are most pugnacious and determined fighters, and therefore not suitable for small yards. If they can get no other enemy they will even fight their own shadows.

The chickens fledge late, and have for a long while a bare, wretched appearance. They require a dry, warm temperature, as in youth, before being fully feathered, they are very delicate and highly susceptible of cold and wet.

The Malays are good layers and sitters and after they are full grown, can be kept most anywhere, but on account of their vindictive cruel nature they are by no means desirable to have and my advice is, to have nothing to do with them.

GAME FOWLS.

BLACK BREASTED RED GAME FOWLS.

This noble race has relationship, though now of remote generations, with the Malays. Before we had any of this breed, the inhabitants of several portions of the Malay or Malacca peninsula, and various parts of the East, possessed them, and used them chiefly for the purpose of cock-fighting.

A thorough-bred Game cock of high degree never fails in courage when opposed to one of his own order. And the Game fowl is the only bird put to the test of combat to prove whether he be genuine or not.

There is a generally recognized standard for form and figure, which must not be departed from, whatever variety of color the birds may present. In weight they vary; four pounds eight or ten ounces was the weight aimed at by the breeders for the cockpit, but six pounds is often reached, when two years old; but beyond this weight impurity of blood may be suspected.

The carriage and form of the Game cock are certainly more beautiful than that of any other variety of domestic fowl. The neck is long, strong and gracefully curved; the hackle short and very close; the breast broad; the back short, broad across the shoulders; the whole body very firm and hard, with a perfectly straight breast and back, the latter tapering toward the tail; the wings large and powerful, and carried closely pressed into the sides; the thighs strong, muscular and short, tightly clothed with feathers, and well set forward on the body, so as to be available for fighting; the shanks rather long, strong but not coarse, covered with fine scales; the feet flat and thin, the toes long and spreading, so as to give a good hold on the ground; the hind toe must be set low down, so as to rest flatly on the ground, and not merely touch with the point—a defect which is known as "duck-footed," and renders the bird unsteady when pushed backward by his opponent.

The plumage is compact, hard and mail-like to a remarkable degree, and possesses a brilliant glossiness that cannot be surpassed. The tail in the cock is rather long, the sickle feathers gracefully arched and carried closely together, the whole tail curved backward and not brought forward over the back—a defect called squirrel-tailed.

The head is extremely beautiful, being thin and long, like that of a greyhound; the beak massive at its root, strong, and well curved; the eye large, very full, and brilliant in lustre; the ear-lobe and face of a bright scarlet, and the comb in undubbed birds single, erect, and thin. The spur, which is exceedingly dense and sharp, should be set low on the leg, increasing its power; spurs are frequently on the hens.

In the hen, the form, making due allowance for the difference

GAME FOWLS.

of sex and alteration of plumage, resembles that of the cock. The head is neater, the face lean and thin. The small thin comb should be low in front, evenly serrated, and perfectly erect.

The deaf-ear and wattles should be small. The neck, from the absence of hackle feathers, looks longer and more slender than that of her mate. The tail feathers should be held closely together, and not spread out like a fan. The plumage should be so close that the form of the wing should be distinctly visible, the outline not being hidden by the feathers of the body.

As the Game fowl is impatient of restraint, a good grass run is essential to keep it in good condition. In breeding great care must be taken in matching, as regards form, feather and the color of the beak and legs. Much depends upon the purity of the hens, for a good Game hen, with a dunghill cock, will breed good fighting birds, but the best Game cock, with a dunghill hen, will not breed a bird good for anything. It is not desirable to mate old birds; a stag, or last year's bird placed with hens two or three years old, will produce finer chickens than when an old cock is mated with last season's hens. For great excellence, four hens with one cock is sufficient.

The hens are good layers and as sitters have no superiors. Quiet on their eggs, regular in coming off, and confident, in their fearlessness, of repelling intruders, they rarely fail to rear good broods, and defend them from violent attacks.

The newly-hatched chickens are very attractive; those of the darker breeds are light brown, with a dark brown stripe down the back and a narrower line over the eye. The duck-wings, grays and blues have proportionally paler hues, but the stripe is seldom absent.

The chickens feather rapidly, and with good care and liberal, varied diet, such as cottage cheese, chopped egg, with a portion of onions, bread crumbs, grits, boiled oatmeal, barley and wheat, with some milk in the earlier stages of their growth, are reared with less difficulty than other fowls.

As Game fowls will fight, and as they are frequently trained for fighting, it is argued that their combs, ear-lobes and wattles should be removed, or "dubbed." This had best be entrusted to the skilled professional.

VARIETIES.

The recognized varieties of Game fowls are—the Black; Black breasted Red; White; White Pile; Blue; Brown-red; Red Pile; Gray; Spangled; Ginger-red; Silver Duck-wing; Yellow Duck-wing.

DOMINIQUE FOWL.

THE DOMINIQUE FOWL.

This seems to be a tolerable distinct and permanent variety, about the size of the common Dunghill Fowl. Their name is taken from the island of Dominica, from which they are reported to have been imported. Take all in all, they are one of the very best breeds of fowl which we have; and although they do not come in to laying so young as the Spanish, they are far better sitters and nursers. Their combs are generally double, and the wattles are quite small. Their plumage presents, all over, a sort of greenish appearance, from a peculiar arrangement of blue and white feathers, which is the chief characteristic of the variety; although, in some specimens, the plumage is gray in both cock and hen. They are very hardy, healthy, excellent layers and capital sitters. No fowl have better stood the tests of mixing without deteriorating than the pure Dominique.

SEBRIGHT BANTAM.

THE BANTAMS.

Bantams are generally kept more for show and amusement than anything else, although, even as profitable poultry they are not destitute of merit; in proportion to the food they consume, they furnish a fair supply of eggs. As table fowls, the hardy little Game Bantams are excellent, plump, full chested and meaty. As useful and ornamental pets, I know of no birds that are superior. The Sebright Bantams are the most esteemed by fanciers. The cocks should not weigh more than twenty-seven ounces; hens about twenty-three, but the lighter in weight the more they are appreciated.

The chicks of the Bantams generally should be hatched in fine weather, and kept for some time in a cozy place.

VARIETIES.

Golden Sebright; Silver Sebright; Game; Rose-combed Black; Rose-combed White; Japanese; Pekin; Booted White; and White-crested White Polish.

GOLDEN SPANGLED HAMBURG COCK AND HEN.

THE HAMBURG FOWL.

These fowls are "Everlasting layers" and are seldom inclined to sit. They are too small in size to rear for table, and I think too delicate when young to rear at all; only they are such wonderfully good layers, that one dislikes to dispense with them. They are also known as Chittaprats, Bolton Greys, Pencilled Dutch, Silver Hamburgs, Creole, Bolton Bays, Golden Hamburgs. They are a very noisy fowl, and if the hen-roost should be disturbed at night, nothing but death or liberty will induce them to hold the peace.

SILVER SPANGLED HAMBURGS.

DESCRIPTION AND VARIETIES.

The Hamburgs have a graceful and upright carriage. The head in the cock is small; beak of a dark color, medium in size; rose comb of a deep red color not inclining to droop on either side, the top covered with small points and ending in a spike; ear-lobes, white of medium size; wattles, red; neck curved; hackle, large and flowing; body, round; breast, very full; plumage close and glossy; legs rather short. The varieties are—Black; White; Golden Pencilled; Silver Pencilled; Golden Spangled; Silver Spangled.

PLYMOUTH ROCKS.

"If there is a better breed for the farmer, or for those who desire both eggs and chickens, we have failed to find it: although many have been tried and 'found wanting.'"

The great popularity that the Plymouth Rock fowl has attained in so short a time, is without a parallel in the annals of gallina-culture, and no other breed is so highly esteemed in America to-day. It has attained this popularity, too, entirely on its own intrinsic merit, without the eclat of foreign origin, or the outlay of large sums of money in "puffing." As *table fowls*, they have no equal in America; being exceedingly sweet, juicy, fine-grained, tender, and delicate. As *spring chickens*, they are the very best breed, for, added to the excellence of their flesh, they feather early, and mature with remarkable rapidity. As *market fowls*, they are unsurpassed, being large (cocks weigh 9 to 11 pounds, hens 7 to 9), and very plump bod:es, with full breasts, clean, bright yellow legs, and yellow skin; they always command the highest price. As *egg-producers*, they are only excelled by the Leghorn class, and lay more eggs than any other breed that

PLYMOUTH ROCKS. 35

hatches and rears its own young, and can be depended upon for eggs all the year round. Their eggs are also of large size, very rich,

and fine-flavored, from white to redish-brown in color. In hardiness, both as chicks and mature fowls, they are also unequaled, and being out-and-out an American breed, they adapt themselves to all climates and situations better than any other breed. Their combs and wattles being of moderate size, are not liable to freeze, and they have no feathers on the lower part of their legs to drabble in the snow and mud, and thereby chill them. In plumage, they are bluish-gray, each feather distinctly penciled across with bars of a darker color, hence are very admirable, and not likely to become soiled by the smoke and dust of the city. Added to their fine plumage, their symmetrical form and upright and pleasing carriage enable them to vie with most breeds, either upon the lawn, in the yard of the fancier, or in the exhibition hall. As mothers, they are excellent, being neither non-sitters nor persistent sitters, are kind and gentle, and good foragers. In disposition, they are quiet, gentle, and cheerful, bear confinement well, and are easily confined, their wings being too small, and bodies too large to admit to much progress in flight. If given range, they will find their own living, and if confined, need a remarkably small amount of food for such large fowls. In fine, this comparatively new breed combines all the sturdy and excellent qualities of the ideal fowl to a wonderful degree, (the merits of the large flesh-producing and small egg-producing breeds,) filling a place long sought for, but never before attained, and is a golden mean. It is pre-eminently the farmer's and mechanics' fowl—in fact the BEST fowl for all who have facilities for keeping but one variety, and desire that one to be a "general purpose" breed.

LANGSHANS.

LANGSHAN FOWL.

The Langshan is the latest acquisition to our poultry yards from Asia, and, judging from our experience with other Asiatic breeds, their origin certainly augers well for their future in this country. They are natives of northern China, and consequently accustomed to its rugged climate.

The discoverer of this variety in China was a scientist in the employ of the British government, and not a "chicken fancier," particularly. Eight years ago, he wrote thus to his English friends: "I send you some fine fowls by the steamer *Archilles*, of Hall & Holt's line. They are clear black, and are called *Langshans*. Look out for their arrival and send for them without delay." * * * A second letter stated that "the fowls I am sending you are very fine. Their plumage is of a bright glossy black. I have never seen any like them before, and I am told their flesh is excellent. The Chinese say they are allied to the wild turkey; they are very valuable birds. You must be very careful of them, and get them acclimated by degrees."

These birds we sent to Major A. C. Croad, Durington Worthing, England, from his nephew, who was, a few years ago, upon an exploring expedition under orders from the English government, in the north of China, where he discovered this fine variety of fowls, in the province of *Langshan*, and sent home the first that were ever seen in England.

Upon the arrival of the *Archilles*, in England, Major Croad lost no time in sending for his birds; and the messenger, on his return, informed him that the new arrivals had received quite an ovation in the docks, people crowding to have a look at them, asking what breed they were, and whether they were for sale, etc. The captain of the steamer told him that, although he had been several times to china, he had never met with any fowls like these before.

The Langshans were publicly exhibited the next year at the Crystal Palace and other leading shows, and were bred successfully for three or four years, the stock being kept under the supervision of the agents of the original importer.

They were of late years imported to America, and our American fanciers speak well of them; in fact they are the best birds that were ever imported from China. Langshans have straight red combs, somewhat larger than those of Cochins. Their breast is full, broad and round, and carried well forward, being well meated, similar to the Dorkings. Their body is round and deep like the Brahmas. The universal color of the plumage is a rich metallic black. The tail is long, full feathered, and of the same color as the body. The color of their legs is a blue black, with a purplish tint between the toes. The average weight of a cockerel, at seven or eight months, when fattened, is about ten pounds; and a pullet about eight pounds. Their carriage is stylish and stately.

The good qualities claimed for the Langshans are the following: They are hardy, withstanding readily even severest weather. They attain maturity quite as early as any of the large breeds. They lay large, rich eggs all the year round, and are not inveterate sitters. Being of large size, with white flesh and skin, they make an excellent table fowl; more especially so on account of the delicacy of the flavor which the flesh possesses. To briefly summarize, I may then say that this breed is worth the attention of all. Firstly, because they come from a part of the world which has given us many of our most excellent breeds; and secondly, because their popularity is in the ascendency, and they seem to combine in themselves nearly all the valuable characteristics that go to make up a practically useful fowl.

I give in connection with this article a wood-cut of a pair of Langshans, believing that a faithful illustration will do more to give an accurate idea than even an extended description. It will be observed that, apparently, they are more like the Black Cochin than any other breed with which we are familiar, but in reality they differ very essentially from them.

WYANDOTTES.

This new breed have so many points to recommend them, both to the fancier and farmer, that they will surely become very popular. Their plumage is white, heavily laced with black, the tail alone being solid black; the lacing on the breast is peculiarly handsome. They have a small rose comb, close-

THE WYANDOTTES.

fitting; face and ear lobes bright red. Their legs are free from feathers and are of a rich yellow color. In shape they bear more resemblance to the Dorkings than any other breed. Hens weigh 8 to 9 pounds, cocks 9 to 10 pounds, when full grown. They are very hardy, mature early, and are ready to market at any age. Their flesh is very fine flavored and close grained, which, with their yellow skin, model shape and fine, plump appearance, particularly adapts them for market. They are extraordinary layers, surprising every breeder at the quantity of eggs they produce. If allowed to sit they make most careful mothers, are content anywhere, and will not attempt to fly over a fence four feet high. Their great beauty and good qualities will make for them a host of friends wherever the breed is introduced.

THE LEGHORN FOWL.

ROSE COMB BROWN LEGHORN.

LEGHORN FOWL.

This abmirable breed of fowls has become widely disseminated in the United States. They are valued for their many good qualities, among which are beauty and constant laying propensities.

BROWN LEGHORN COCK.

They are very hardy fowls, possessing all the advantages of the Spanish without their drawbacks. Their legs are bright yellow, and perfectly free from feathering on the shanks. The faces are red, the ear lobes only being white. The comb in the cock is thin, erect and evenly serrated. In the hen it falls over like that of a Spanish hen. The tail in the cock is exceedingly well furnished with side sickle-feathers, and in both sexes is carried perfectly erect. The birds are active, good foragers, and have a very handsome and sprightly carriage. They are abundant layers of full sized eggs, the hens rarely showing any inclination to sit, but laying the whole year round, except during the annual month. The chickens are very hardy; they feather quickly and mature rapidly, thus having the advantage over the Spanish.

BROWN LEGHORN HEN.

These fowls are exceedingly useful as well as ornamental addition to our stock of poultry; they are more valuable to egg-farmers than breeders of table fowls, as they are but small eaters and so do not put on flesh quickly. To people, however, who depend on their poultry bringing them a constant supply of eggs, they are invaluable.

LEGHORN VARIETIES.

Black; White; Brown, and Dominiques.

LEGHORN FOWL.

THE FRENCH BREEDS.

CREVECŒURS.—These birds are generally supposed to be of Norman origin, and to owe their name to the little village of Crevecœurs, not far from Lisieux. They are fine, well plumaged black birds, with large crests on their heads, in the front of which are situated the two horns, or spikes, which arise from the bifurcation of the comb. They give the bird a very curious look, and make his head resemble the pictures of that of his Satanic majesty. The birds are well shaped, with rather large legs of a leaden grey color. The hens lay large white eggs, but are not good sitters. The pullets mature early, and as they lay soon, put on fat readily, and are of a good shape for table; they are, in dry warm localities, profitable fowls to keep; they bear confinement well, but are rather difficult to rear, and have a decided tendency to "roup.,' If crossed with Brahmas or Leghorns they might probably become more hardy.

LA FLECHE FOWLS.

These birds may be considered, I think, the best of the French fowls for table; they are also more hardy than the Crevecœurs, and have more size and more style, being handsome, upstanding birds, in color jet black, with rich, metallic plumage; their ear lobes are large and perfectly white, their faces bright red and free from feathers. The comb in good well-bred birds does not vary with the sex, and is in the shape of a pair of straight horns; the leg-scales are lead color, hard and firm. The cocks are tall without being at all leggy; the hens have large and rather long bodies, longish necks, and thin clean legs. The best specimens come from the North of France, though they are not even there easy to procure, as the French do not go in for keeping the different breeds of fowls distinct, so it is hard to obtain really pure-bred birds.

HOUDANS.

HOUDAN COCK AND HEN.

These are considered the best French fowls, and of late years have become great favorites with poultry-fanciers. They have, like the Dorkings, five claws on each foot; their plumage is black and white, shaded with violet and green; they are crested birds, the crest turning backwards over the neck; their cheeks are well feathered, and wattles well developed. They differ from other species by several remarkable traits, the head forms a very obtuse angle with the neck, so that the beak is depressed and viewed from above appears like a nose. The flat square comb looks like a fleshy forehead; the cheeks are surrounded with curling feathers which resemble whiskers; the reversed corners of the beak have the appearance of a mouth. The crest looks like a head of hair, and the entire visage instantly reminds the spectator of a man's face.

Houdans are hardy, not difficult to rear good steady layers, but non-sitters; they put on fat readily, and are very good table fowls, flesh excellent and shapely in form.

THE DOMESTIC TURKEY.

DOMESTIC TURKEY.

The domestic turkey can scarcely be said to be divided, like the common fowl, into distinct breeds; although there is considerable variation in color, as well as in size. The finest and strongest birds are those of a bronzed-black; these are not only reared the most easily, but are generally the largest, and fatten the most rapidly. Some turkeys are of a coppery tint, some of a delicate fawn-color, while others are parti-colored, grey, and white, and some few of a pure snow white. All of the latter are regarded as inferior to the black, their color indicating something like degeneracy of constitution.

To describe the domestic turkey is superfluous; the voice of the male; the changing colors of the skin of the head and neck; his proud strut, with expanded tail and lowered wings, jarring on the ground; his irascibility, which is readily excited by red or scarlet colors, are points with which all who dwell in the country are conversant.

The adult turkey, is extremely hardy, and bears the rigors of winter with impunity even in the open air; for during the severest weather, flocks will frequently roost at night upon the roof of a barn, or the branches of tall trees, preferring such an accommodation to an indoor roost. The impatience of restraint and restlessness of the turkey, render it unfit company for fowls in their domitory; in fact the fowl house is altogether an improper place for these large birds, which require open sheds and high perches, and altogether as much freedom as is consistent with their safety.

Although, turkeys will roost even during the winter months on trees, it is by no means recommended that this should be allowed, as the feet of these birds are apt to become frostbitten from such exposure to the air on the sudden decline of the temperature far below the freezing point.

Turkeys are fond of wandering about pastures, and the borders of fields, or in fact any place where they can find insects, snails, slugs, etc., which they greedily devour. In the morning, they should have a good supply of grain, and after their rerurn from their peregrinations another feed; by this plan, not only will the due return home of the flock be insured, but the birds will be kept in good condition, and ready at any time to be put upon fattening diet. Never let them be in poor condition—this is an axiom in the treatment of all poultry—it is difficult, and takes a long time, to bring a bird into proper condition, which has been previously poorly fed or half starved.

The turkey hen is a steady sitter; nothing will induce her to leave the nest; indeed, she often requires to be removed to her food, so overpowering is her instinctive affection; she must be freely supplied with water within her reach; should she lay any eggs after she has commenced incubation, these should be removed—it is proper, therefore, to mark those which were given to her to sit upon. The hen should now on no account, be rashly disturbed; no one except the person to whom she is accustomed, and from whom she receives her food, should be allowed to go near her, and the eggs, unless circumstances imperatively require it, should not be meddled with.

The hen usually sits twice in the year, after laying from a dozen to fifteen or more eggs, on alternate days, or two days in succession, with the interval of one day afterwards, before each breeding. She commences her first laying in March; and if a second early laying is desired, after she has hatched her brood, it is economical to transfer the chicks immediately after they leave the shell to another turkey-hen which had begun to incubate contemporaneously with her, and will now take willing charge of the two young families. This, however, cannot be viewed as a benevolent proceeding; and much less so if the mother be deprived of her offspring, and the consequent pleasure of rearing them, for the purpose of putting a fresh set of eggs under her, which she will steadily hatch for three or four weeks more. In this case, however, fowls' eggs are usually given, from merciful consideration to abridge the period of incubation from thirty-one to twenty-one days.

According to the size of the hen, the season, and the range local temperature, the number of eggs for each hatch may be stated at from eleven to seventeen; thirteen is a fair average number. As the hen lays them, her eggs should be immediately removed, and kept apart until the time for sitting them; else the awkward bird might break them in the nest, as she goes in or out of it. While she is incubating, the cock bird should not be permitted to approach it, lest he should mischievously break the eggs or disturb the hen.

On about the thirtieth day, the chicks leave the eggs; the little ones for some hours will be in no hurry to eat; but when they do begin, supply them constantly and abundantly with chopped eggs, shreds of meat and fat, curd, boiled rice, mixed with lettuce, and the green of onions. Melted mutton suet poured over barley or Indian-meal dough, and cut up when cold is an excellent thing. Little turkeys do not like their food to be minced much smaller than they can swallow it; indolently preferring to make a meal at three or four mouthfuls to troubling themselves with the incessant pecking and scratching in which chickens so much delight. But at any rate, the quantity consumed costs but little; the attention to supply it is everything.

As in the case of young fowls, the turkey chicks do not require food for several hours after they have emerged from their shells.

It is useless to cram them as some do, fearing lest they should starve; and besides, the beak is as yet so tender that it runs a chance of being injured by the process. There is no occasion for alarm if, for thirty hours, they content themselves with the warmth of their parent and enjoy her care. When the chicks feel an inclination for food, it will soon become apparent to you by their actions, then feed them as I have before directed.

FATTENING.

About the middle of September or the first of October, it will be time to begin to think of fattening some of the earliest broods, in order to supply the markets. A hen will be four or five weeks in fatting; a large cock two months or longer, in reaching his full weight. The best diet is barley or Indian meal, mixed with water, given in troughs that have a flat board over them, to keep dirt from falling in. A turnip with the leaves attached, or a hearted cabbage, may now and then be thrown down to amuse them. When they have arrived at the desired degree of fatness, those which are not wanted for immediate use must have no more food given them than is just sufficient to keep them in that state; otherwise the flesh will become red and inflamed, and of course less palatable and wholesome. But with the very best management, after having attained their acme of fattening, they will frequently descend again, and that so quickly, and without apparent cause, as to become quite thin. Turkeys fatten faster, and with less expense, by caponizing them, which, also, produces better and sweeter flesh.

THE GUINEA FOWL.

THE GUINEA FOWL.

Of all known birds, this, perhaps is the most prolific of eggs. Week after week and month after month see little or no intermission of the daily deposit. Even the process of moulting is sometimes insufficient to draw off the nutriment the creature takes to make feathers instead of eggs. From their great aptitude for laying, and also from the very little disposition they show to sit, it is believed, that these birds in their native country, (Africa) do not sit at all on their eggs, but leave them to be hatched by the sun.

It is not every one who knows a cock from a hen of this species. An unerring rule is, that the hen alone uses the call note "come back," "come back," accenting the second syllable strongly. The cock has only the harsh shrill cry of alarm, which, however, is also common to the female.

There is one circumstance, in regard to the habits of the guinea cock, that is, he pairs only with his mate in most cases, like a partridge or a pigeon. In the case where a guinea cock and two hens are kept, it will be found, on close observation, that though the three keep together so as to form one pack, yet that the cock and one hen will be unkind and stingy to the other unfortunate female, keep her at a certain distance, merely suffering her society. The neglected hen will lay eggs, in appearance, like those of the other, in the same nest. If they are to be eaten, all well and good; but if a brood is wanted and the eggs of the despised one chance to be taken for the purpose of hatching, the result is disappointment and addled eggs.

It is best to hatch the eggs of the guinea fowl under a hen of some other species; a Bantam hen makes a first class mother, being lighter, and less likely to injure the eggs by treading on them than a full sized fowl. She will well cover nine eggs, and incubation will last about a month.

Feed the chicks frequently, five or six times a day is not too often, they have such extraordinary powers of digestion, and their growth is so rapid, that they require food every two hours. A check once received can never be recovered. In such cases they do not mope and pine, for a day or two, like young turkeys under similar circumstances and then die; but in half an hour after, being in apparent health, they fall on their backs, give a convulsive kick or two, and fall victims to starvation. Hard-boiled egg, chopped fine, small worms, bread crumbs, chopped meat, or suet, whatever, in short, is most nutritious, is their most appropriate food.

THE DOMESTIC GOOSE.

THE DOMESTIC GOOSE.

With respect to the range and accommodation of geese, they require a house apart from other fowls, and a green pasture, with a convenient pond or stream of water attached. The house must be situated in a dry place, for geese at all times, are fond of a clean, dry place to sleep in, however much they may like to swim in water. It is not a good method to keep geese with other poultry; for when confined in the poultry-yard, they become very pugnacious, and will very much harrass the hens and turkeys.

In allowing geese to range at large, it is well to know that they are very destructive to all garden and farm crops, as well as to young trees, and must, therefore, be carefully excluded from orchards and cultivated fields. It is usual to prevent them getting through the gaps in fences, by hanging a stick or "yoke" across their breast.

Those who breed geese, generally assign one gander to four or five females. When well fed, in a mild climate, geese will lay twice or three times a year, from five to twelve eggs each time, and some more, that is, when they are left to their own way; but if the eggs be carefully removed as soon as laid, they may be made, by abundant feeding, to lay from twenty to fifty eggs without intermitting. They begin to lay early in the spring, usually in March, and it may be known when an individual is about to lay, by her carrying about straws to form her nest with; but, sometimes, she will only throw them about.

When a goose is observed to keep her nest longer than usual, after laying an egg, it is a pretty sure indication that she is desirous of sitting. The nest for hatching should be made of clean straw, lined with hay, and from fourteen to eighteen eggs will be as many as a large goose can conveniently cover. She sits about one month, and requires to have food and water placed near her, that she may not be so long absent as to allow the eggs to cool. The most economical way of getting a great number of goslings, is to employ turkey hens to hatch, and keeping the goose well fed she will continue laying.

Goslings must be kept from cold and rain as much as possible. Feed them on barley or Indian meal or crusts of bread soaked in milk.

VARIETIES.

African; Toulouse; Embden; Egyptian; White Chinese; Brown Chinese.

THE DUCK.

MUSCOVY DUCK.

It is not in all situations that Ducks can be kept with advantage; they require water much more, even, than the goose; they are no grazers, yet they are hearty feeders. Nothing comes amiss to them in the way of food: green vegatables; kitchen scraps; meal of all sorts made into a paste; grains; bread; worms; insects; all are accepted with eagerness. Their appetite is not at all fastidious; in fact they eat most everything, and eat all they can. They never need cramming, give them enough, and they will cram themselves; but remember, confinement will not do for them; they must have room, and plenty of it, also a large pond or stream, if you have these requirements they can be kept at little expense.

Where they have much extent of water or shrubbery to roam over, they should be looked after and driven home at night, and provided with proper houses or pens; otherwise they are liable to lay and sit abroad. As they usually lay either at night, or very early in the morning, it is a good way to secure their eggs, to confine them during the period when they must lay, a circumstance easily ascertained by feeling the vent.

COMMON DUCK.

The duck is not naturally disposed to incubate, but in order to induce her to do so, you may, towards the end of the laying, leave two or three eggs in the nest, taking care every morning to take away the oldest laid, that they may not be spoiled. When she shows a desire to sit, from eight to ten eggs may be given according to the size of the duck, and her ability to cover them. The duck requires some care when she sits; for as she cannot go to her food, attention must be paid to place it before her; and she will be content with it, whatever be its quality; it has been remarked that when ducks are too well fed, they will not sit well. The period of incubation is about thirty days.

WILD DUCK.

The duck is apt to let her eggs get cold, when she hatches and many thereby are lost, this together with the fact of her often leading the ducklings into the water immediately after they are excluded from the shell and thus losing many if the weather is cold, often induces poultry keepers to have duck eggs hatched by hens or turkey hens; and being more assiduous than ducks, these borrowed mothers take an affection for the young, to watch over, which requires great attention because as these are unable to accompany them on the water, for which they show the greatest propensity as soon as they are excluded, they follow the mother hen on dry land, and get a little hardy before they are allowed to take to the water without any guide.

The best mode of rearing ducklings depends very much upon the situation in which they are hatched. For the first month, the confinement of their mother, under a coop is better than too much liberty. All kinds of sopped food, buckwheat flour, Indian or barley meal and water mixed thin, worms, &c., suit them.

When ducklings have been hatched under a common hen, or a turkey hen and have at last been allowed to go into the water, it is necessary, to prevent accidents, to take care that such ducklings come regularly home every evening; but precautions must be taken before they are permitted to mingle with the old ducks lest the latter ill-treat and kill them, though ducks are by no means so pugnacious and jealous of new-comers as common fowls uniformly are.

VARIETIES.

ROUEN DUCKS.—The flesh is abundant and of good flavor; good specimens will dress from five to seven pounds each.

AYLESBURY DUCKS.— These are considered the most valuable of the English breeds and is well thought of in this country. They are good layers, but do not weigh quite as much as the Rouen breed.

CAYUGA DUCKS—These are the finest of the American breeds, they are also the largest and most valuable of the duck family. They weigh generally from eight to ten pounds, are good layers, and easily raised.

The other varieties are the Mandarin; Carolina; Muscovy; Call Duck; Black East India.

The duck is peculiarly the poor man's bird (its hardihood renders it so entirely independant of that care which fowls perpetually require); and indeed of all those classes of persons in humble life, who have sloppy offal of some sort left from their meals, and who do not keep a pig to consume it. Ducks are the best save-waste for them; even the refuse of potatoes, or any other vegetables will satisfy a duck, which thankfully accepts, and with a degree of good virtue which it is pleasant to contemplate, swallows whatever is presented to it, and very rarely occasions trouble. Though fowls must be provided with a roof and a decent habitation, and supplied with corn, which is costly, the cottage garden waste, and the snails and slugs which are generated there, with the kitchen scraps and offal, furnish the hardy ducks with the means of subsistence. And at night they require no better lodgings than a nook in an open shed; if a house be expressly made for them, it need not necessarily be more than a few feet in height, nor of better materials than rough boards and clay mortar, a door being useless, unless to secure them from thieves.

POINTS OF POULTRY.

A—Neck hackle. *B*—Saddle hackle. *C*—Tail. *D*—Breast. *E*—Upper Wing coverts. *F*—Lower Wing coverts. *G*—Primary quills. *H*—Thighs. *I*—Legs. *K*—Comb. *L*—Wattles. *M*—Ear-lobe.

Dictionary of Poultry Terms.

BEARD.—A bunch of feathers under the throat of some breeds, as Houdans or Polish.

BREED.—Any variety of fowl presenting distinct characteristics.

BROOD.—The number of birds hatched at once; a family of young chickens.

BROODY.—When the hen desires to sit she is said to be broody.

CARRIAGE.—The upright attitude or bearing of a fowl.

CARUNCULATED.—Having a fleshy excrescence or protuberances, as on the neck of a turkey-cock.

CHICK.—A very young fowl.

CHICKEN.—A name applied to fowls until they are full grown.

CLUTCH.—The eggs placed under a sitting hen, also the brood hatched therefrom.

COCKEREL.—A young cock.

COCK.—The full grown male bird.

COMB.—The crest or red fleshy tuft growing on top of a fowl's head.

CREST.—A top-knot of feathers, as on the head of the Polands.

CROP.—The first stomach of a fowl, through which the food must pass before the process of digestion begins.

DEAF-EARS.—Folds of skin hanging from the true ears, varying in color.

DUBBING.—To cut off the comb, wattles, &c., leaving the head smooth.

EAR-LOBES.—Folds of skin hanging from the ears.

FACE.—The bare skin extending from the top of the bill around the eyes.

FLIGHT-FEATHERS.—The primary wing feathers, used in flying.

FLUFFS.—The downy feathers around the thighs.

GILLS.—A term sometimes applied to the wattles; the flap that hangs below the beak.

HACKLES.—The peculiar narrow feathers on a fowl's neck.

HEN-FEATHERED.—A cock, which owing to the absence of sickle feathers resembles a hen.

HENNY.—The same as hen-feathered.

HOCK.—The elbow joint of the leg.

KEEL.—The breast bone.

LEG.—The shank.

LEG-FEATHERS.—Feathers growing on the outside of the shank.

MOSSY.—Uncertain marking.

PEA-COMB.—A tripple comb.

PENCILING.—Small stripes running over a feather.

POULT.—A young turkey.

PRIMARIES.—The same as flight-feathers.

PULLET.—A young hen.

ROOSTER.—A word used in the United States to designate the male fowl; generally called cock.

SADDLE.—The posterior of the back, the feathers that cover it are termed saddle-feathers.

SECONDARIES.—The quill-feathers of the wing, which show when the fowl is at rest.

SHANK.—The leg.

SICKLE-FEATHERS.—The upward curving feathers of a cock's tail.

SPANGLED.—Spots on each feather of a different color from that of the ground color of the feather.

SPUR.—A sharp bone protruding from the heel of a cock.

STRAIN.—A race of fowls that has been bred for years unmixed with other breeds.

TAIL-COVERTS.—The curved feathers at the sides of the bottom of the tail.

TAIL-FEATHERS.—The straight feathers of the tail.

THIGHS.—The upper part of the shanks.

TOP-KNOT.—The same meaning as crest.

TRIO.—One cock and two hens.

VULTURE-HOCK.—Stiff projecting feathers at the hock-joint.

WATTLES.—The red fleshy excrescence that grows under the throat of a cock or a turkey.

WING-BAR.—A dark line across the middle of the wing.

WING-COVERTS.—The feathers covering the roots of the secondary quills.

POULTRY-KEEPING.

Any person who takes up poultry-keeping should have *some end* in view; should either keep fowls for showing and prize-taking, or for laying and fattening. Fowls for domestic use and fowls for exhibition are two totally different things, and call for entirely different methods of treatment.

In this small book I wish to adhere as much as possible to the business of poultry-keeping on *a small scale* within the means of all people living in the country, and having a little ground of their own.

If there is a farm-yard to fall back on, and the birds are not kept by themselves, but are allowed to run with the other inmates of that yard, having a hen-house in which to roost, lay, and sit, then your cares are reduced to a minimum. As all who might and should keep poultry, have, however, no farm, but only a garden and a plot of ground, I will not say any more about the old farmyard system, but suppose that the fowls have to be kept on a small scale without the foregoing advantages. Much depends on the purpose for which fowls are kept, if for show and prize-taking, or merely for domestic uses, for table and for eggs.

If for show, then the different breeds must be kept thoroughly pure, entirely distinct, and great attention given to points generally. A higher class of fowl must be purchased in the first instance; the diet must be more generous, size being a great point with judges; and the whole business of poultry-keeping is placed on a more costly footing, and becomes an expensive and but rarely a remunerative amusement; whereas in merely keeping a small stock of fowls for table use—the first and original outlay of purchase and house-building overcome—you should, and can easily, have, with a little trouble, a small profit each month after the necessary food is paid for.

I have done both myself: kept fowls for general use—ordinary common birds, mostly cross-bred—and kept purely-bred birds to show, and I have no hesitation in saying that the former is the best plan, unless, of course, you are a poultry fancier and have money enough to allow you to indulge your mania for prize birds; then, with highly-bred stock, you may look to the sale of eggs and the taking of prizes to, in a measure, recoup you for your outlay. I was fairly successful with the high-class birds I purchased, and got good prices for the sitting of eggs I sold, as also for the birds themselves when I parted with them; but I cannot honestly say I consider the keeping up of select and distinct sorts is worth the trouble it entails—that is, if you do the work of looking after them yourself. Mine, I know—I could not afford to keep a poultry-man — led me a sad dance. I was always in trouble with them; they had separate houses and runs, but unless I was near while the different sorts were having their outing there was sure to be some disturbance, a fight between the cocks through the bars and netting, and this very likely occurred just before I wanted to show one of them, when featherless heads and wounded bleeding combs would be the result; and the hens too were nearly as pugilistic. Some one will probably remark, "Mismanagement." Possibly; but I had not all the proper arrangements a regular prize poultry-breeder would have, and even in the very best regulated poultry-yards accidents will, we know, occur, and so these creatures were a perpetual torment to me.

And when, after an interval of some years, I began poultry-keeping again, I started on an entirely different plan and on a very small scale. It is from my experience then gained that I offer the following hints to those living in the country who wish to keep poultry and yet do not mean to incur much expense in so doing.

For general use I would say do not keep entirely to pure-bred birds, but mix them with others; a good cross-breed is often more desirable than a really pure breed; not only are the fowls resulting from the cross stronger and less likely to become sickly and degenerate, but you can, by a judicious selection in the cross you allow, counteract many of the qualities you do not consider quite desirable.

BEST BREEDS FOR MARKET.

I do not believe there are any better market fowl, all things considered than the Langshans, next comes the Brahmas. The Dorkings are a superior table fowl, but are tender and hard to rear.

EGG PRODUCERS.

The Black Spanish; Polands; Houdans and Hamburgs are all inveterate layers; but the Black Spanish and Hamburgs are rather tender, and more fit for the fancier than for the practical man. For a desirable "all round" breed, I should recommend the Plymouth Rocks. At any rate I have described the different breeds, given you their good and bad points, and you may take your choice.

SORTS FOR SMALL YARDS.

If you have only a limited space to allow for your birds, do not keep too many at first. Possibly, as you find your poultry answer, you may wish to considerably increase your stock, and so will have to enlarge your premises, which by that time you may be able to do; besides, you will have gained experience during the time you have been looking after a limited number, and will have learned many things respecting the nature of fowls, their habits, diseases, constitutions, and general characteristics, of which before you were entirely ignorant.

My own opinion is, I own, entirely against a very large poultry-farm. I should always prefer having a small one under my own immediate eye to possessing a quantity of birds and being obliged to keep a man or woman to look after them.

If people *want to lose money by poultry* let them mass them in numbers, and they will soon gain the desired result. If, on the contrary, they will be content with modest profits, and patiently turn over pennies instead of expecting to turn over dollars, then let them keep poultry on a small scale, attend to them themselevs, spare no pains or trouble in looking after and thoroughly understanding the requirements of their stock, and they need not fear but that the result will be satisfactory.

To buy pens, nests, rent land, pay a man to look after the stock, waste money in sundries and expensive food, buy useless items, and hand over all trouble to subordinates, *is not the way to make poultry pay.*

While, on the other hand, to look after a little poultry-yard yourself, to vary the food by economising all kitchen refuse, buying up cargo rice and second-class grain—which is really quite good enough for fowls, and better suited to them than very good barley, oats, or wheat—never to allow food to be wasted, nor to keep an old and useless stock, *is the way to insure certain small profits,* if those will content you.

In trying to grasp too much you stand a chance of losing even more than the original outlay. A great many people, who have now little plots of ground suitable for fowls, but standing empty, are deterred from keeping poultry by the idea that it is so expensive a proceeding, and that they will eventually be out of pocket by it. So they will, certainly, if they commence on too large a scale; but if they began with a dozen or two dozen fowls, and and kept the original stock down to that number, only allowing the chickens for killing during the season, and pullets for laying to swell the numbers each year, then we should hear less about poultry expenses, and more about eggs and chickens.

Now with regard to commencing operations. BRAHMAS, LEGHORNS, PLYMOUTH ROCKS and LANGSHANS are the fowls I should keep. Brahmas as winter layers, good sitters, and good mothers; Plymouth Rocks as good all round, and Langshans are especially for table.

The number of hens I would allow to each cock would be: Leghorns, twelve hens to each cock; Brahmas, eight hens to one cock; Plymouth Rocks, six hens to one cock; Langshans, six hens to one cock.

THE FIRST OUTLAY.

If you have an adaptable outhouse, which can, with a little contrivance and a little money spent on it, be turned into a fowl-house, you are indeed lucky, for you will then for a few dollars, say fifteen at the outside, be able to fit it with perches and nests, and see to the flooring, roofing and ventilation.

Your nests, of strong wickerwork or straw, will not cost you more than 25 cents each. You should have twelve at first. You can easily have more if you want them for sitting purposes, but you certainly will not require a nest for each hen. An old saucepan for cooking the food your kitchen will probably supply. Your water-pans should be of common strong yellow stoneware.

If you have no run, you must inclose one with wire, and this will be rather expensive; but your fowls, if they are to be kept in a certain degree of confinement, must have exercise, so a run' or yard is an *absolute necessity*.

You should have a door in the run, at one of the ends adjoining the house, and a door besides in the house itself, with an opening in it, closed by a slide, for the fowls to go in and out as they like.

In the run must be the sheltered place for the dust-bath and for the birds to run under in case of rain. (*See* "Houses an.

FIRST OUTLAY. 71

Yards.") This inclosure I should not have covered at the top. The height of three rows of wire, one on top of the other—*i. e.*, 72 inches—will be quite high enough to prevent heavy birds getting out over; and the Hamburghs, who are by nature great roamers, must have their wings clipped; it will not improve their appearance, but, as they are not kept for showing, that will not much matter. The wire netting you will be able to fix yourself with a little help, unless you are lone women in the house, in which case you will have to get some man who is clever at doing odds and ends of work to help you.

It is a fatal error to cramp fowls. Better far to have a small healthy family of poultry than a large sickly one. If a few birds are well looked after and made comfortable they will be more likely to pay than a number badly kept and allowed too little room.

If from want of space or want of money you can only keep a few fowls, do not be discouraged. A cock and a couple or three or four hens will not eat much, but on the principle of "every little helps" the eggs and two or three broods of chickens from them in the year will be something; they will give you amusement in looking after them, and if you do not sell but merely eat the eggs and the chickens, they will help out the household bills and pay for the extra food you will require; for with only three or four birds, household scraps, if carefully economised, and a little grain daily, will be quite enough to keep them healthy.

There are a great number of poultry-books, and very excellent ones too; but most of them are written with the object of instructing would-be poultry-keepers in the method of keeping a large number of fowls, but few hints being given to people who can only afford to keep a few, and those not for exhibition and show, but really for use. It is, however just these small poultry-keepers we want to see multiplied in America, for until poultry-keeping becomes a national industry—which it cannot unless taken up by the million—so long will the money which should be kept in the country be sent out of it for eggs and chickens, more particularly for the former articles.

Poultry-farming on a large scale has been tried often in America, of late years more especially, but hitherto it has not proved very successful; it does not do so, though, in other countries. When fowls are massed they become unhealthy; this has been proved very frequently. It is not poultry-farming on an extensive plan, however, that I advocate, but *general fowl-keeping*. I would wish to see every laborer with his few fowls, making a little extra money by the eggs and chickens they produce. To do this, however, profitably there must be *thrift*, and in this valuable quality I fear the Americans as a nation are found wanting. Our cooking is by no means good or economical; this is a well-known fact. Where a French peasant's wife will set her husband down to appetising food, be it only a tasty *potage*, an American mechanic's wife will put before hers ill-cooked food costing far more, but less nourishing from the fact of its being so badly dressed. Here is a decided want of *thrift*. So in poultry-keeping, peasants in France keep a cock and a few hens as a matter of course, but feed them very economically on household and garden scraps, various odds and ends, and so make them not only pay their way, but help in the housekeeping besides. Unfortunately we as a nation do not care to trouble over small matters, or attend to the merest details, as the French do; yet it is just this attention to trifles which makes poultry-keeping on a small scale pay.

It is far better to attend to everything yourself—in fact, unless you have plenty of money and can have an experienced man or woman to look after your stock, you must do so. Leave nothing undone for the comfort of your birds, and go through your daily work in your poultry-yard regularly and methodically.

HOUSES AND YARDS.

If you have to build a fowl-house it need not be in any way an expensive erection. Let it be, if possible, built on to an outside wall of the house, say with its back to the kitchen or greenhouse, in such a position as to insure some degree of warmth to the inmates. Let the floor be dry, the roof weather-tight, and the ventilation good, and your fowls will be sure to do well in it. The cheapest material to make it of would be *rough boards*. The roof can also be boarded, only in that case it should be covered with felt. The holes for ventilation should be so placed that the birds feel no cold air on them while roosting. Such a house should measure at least eight feet square, and the roof should slope from about seven to five feet. The door should lock, and a trap-door should be made in it for the hens to go in and out at will: this trap-door should be a sliding one, and easily closed when required, at night being always kept shut for fear of foxes, cats, &c.

Perches should be round poles, not less than four or five inches in diameter, and should not be set too high up—an error into which many people fall. Three feet from the ground is quite high enough for the most elevated perch, and there should be others lower, two and a half feet and two feet from the ground

If perches are too high, heavy fowls cannot fly up to them with ease, and in descending are certain in time to injure themselves, bending or breaking the breastbone and injuring their feet.

The floor should not be of brick, stone, or wood, but of beaten earth well battened down until it presents a perfectly smooth, hard surface, which should be swept out carefully daily and sanded or sprinkled with fine sifted ashes. If, however, you have to build a house for your birds, there being no outhouse you can turn into a fowl-house, then you might prepare a floor of either chalk battened down until quite hard, the ground being dug out to the depth of a foot and filled in with the chalk, over which should be spread sifted ashes or sand; or else fill in the space dug out with burnt clay, also thoroughly rammed down and spread over with a wet mass of cinders, fine gravel, quicklime, and water; this when dry forms a very good floor.

The nests should be arranged so that they are screened from view and darkened, not placed high up for the same reason as before given with regard to the perches, and they should have a ledge in front of them for the hen to step on before going into her nest or on leaving it, else in flying down eggs are frequently dragged out and broken in the fall; and if chickens are hatched high up they are liable to creep out of the nest, fall down, and die. Soft straw is the best lining for nests, as it does not harbor insects so much as hay. It should be *frequently changed* unless hens are sitting, and then it is best not to disturb the hen, or she may forsake her nest. Nest-eggs of stone or china are easily procured, and should be kept. Many hens will not lay in a nest unless there is an egg already in it, and will forsake a nest they have been laying in if all the eggs are removed. Some people leave in the nest an ordinary egg, but this plan is most objectionable; it imparts to the nest a musty smell, and gives also a taste of *must* to those fresh eggs which are laid in it, and which, though really fresh in themselves, have thus a disagreeable odor and taste, quite leading one to suppose that they were stale. This is the reason why so many eggs brought to table have this defect; people will not take the trouble to change the straw in the nest often enough. Besides all this there is the danger of the stale nest egg breaking, which if it does, the nest, and even the whole hen-house, will become offensive. A stone nest-egg can always be kept in a nest, and if a hen wants to sit, a few placed under her form a good trial of her steady sitting powers, and settle her on her nest before the real eggs she has to hatch out are placed under her.

If you keep more than one sort of fowl you must have divisions in your houses. If it is built either against the kitchen wall, or back to some room in which there is in winter constantly a fire, the effect of the warmth will be apparent in the greater number of eggs your hens will lay during the cold weather. Or the hen-house could be built on to a greenhouse wall which is kept heated in the winter. The nests should be resting against the warm back wall, and the birds roosting on the perches will also feel the benefit of the heat. It is astonishing how much fowls enjoy warmth. This is the reason why cottager's fowls lay often very much earlier than those kept by amateurs, because they are generally kept in a lean-to outhouse built against the cottage wall close to the fireplace. The fowls by this means get the warmth of the fire, and in some cases they actually roost in the kitchen. All poultry-keepers could have their fowl-houses run up outside some fireplace or flue, which would keep the birds warm without the expense of an extra fire.

Yard or Run.—If fowls are not allowed free range, which is not always possible on account of gardens or neighbors, a space should be inclosed for them, either fenced off with wooden pailings or wire netting. In this run should be a plot of grass, and if possible a shrub or two for the birds to pick insects off. If the space allows of it there should be a small covered shed in one corner for the fowls to run under during the rain, as fowls cannot endure damp, and under this shed should be the *dust-bath.* It is a downright necessity for all birds to roll or *bathe* in the dust.

They are very particular about their toilets. This may sound to some absurd, but it is most important. No fowls will keep in health unless they are clean, and by rolling in fine dust and ashes, and covering themselves with them, they clean themselves and get rid of the fleas and parasites with which they are always more or less infested. Fowls that are allowed their entire freedom always make dust-heaps for themselves, and retire to them daily.

If it is possible to have a little running stream conducted through this yard then you may indeed consider yourself fortunate, but most likely you will have to content yourself with pots and pans for water. Let these be shallow, and change the water frequently. The question of coops for chickens I have considered in the chapter on Hatching, but I may mention here that the shed in the yard would be a very good place for mother hen and her family when the weather was damp. A shed need not be an expensive building. A few rough poles, with a felt roof, could be easily made by any one, and it is a very great boon to fowls. It need not be of any great size or height, only the roof should have a considerable slope for the rain to run off.

FOOD.

Overfeeding is as great a mistake as underfeeding. Three times a day is quite enough to feed old fowls: a good meal in the early morning, another before going to roost, and a midday feed. Many people, however, only feed twice; this, if the fowls have a farmyard to dig and forage about in, is enough, but in limited space I should certainly feed three times, giving grain for the last meal as more sustaining and stimulating. Chickens, of course, require food much more frequently.

Before I describe the various sorts of food suitable for poultry, a few general directions will be advisable.

Feed regularly—that is, at stated hours—and do not get into the habit of giving handfuls of grain in and out in the course of the day; if you do so you will spoil the birds' digestion.

A supply of pure fresh water is another absolute necessity. Every day in winter the pans should be washed out and filled with fresh water, twice a day in summer if the weather is very hot.

All poultry like *a change of diet*, and should on no account be fed day after day with the same food; as fowls are not fastidious, but will eat nearly any food, there is no possible reason why a variety of food should not be given them, and it is certain they will thrive and do better when their tastes are consulted a little.

Rice is a cheap food, but is not very nutritious, therefore should be given mixed with other foods; it may, however, be considered as an excellent food for fowls which are not kept up for show purposes, and if poultry are suffering at all from diarrhœa should be at once given instead of their ordinary food. Rice, whenever given, should be *cooked* as the raw grain is most injurious, and by swelling in the crop after it has been swallowed often makes the fowl "crop-bound." It should be prepared thus:—Boiled until the grains are completely separated, not in hard lumps, but easy for the birds to pick up when scattered about in the yard; a piece of dripping dissolved in the water in which the rice is boiled has a wonderfully softening effect on it. In winter I always mix a little coarse black pepper with the cooked rice. Fowls in cold weather need stimulants; and pepper, when given in sparing quantities, is very good for them. Rice can often be purchased very cheaply; many grocers sell what they term "fowl-rice," but if you are tolerably near a seaport you can very often get the chance of buying damaged "cargo" rice, which, though possibly just a little injured by seawater, is still excellent food for poultry. I have ranked rice first because of its cheapness. Of the different sorts of corn *barley* is the least expensive, but it is too heating to feed fowls on it alone; it should be ground into meal, mixed with water and fine bran or scraps, and given in a crumbly state, not too moist nor yet too lumpy.

Cooked or prepared food is good for all live stock of all descriptions, for experience proves it to be more nutritious from the changes effected, and therefore more readily digested. One writer advises the following mixture:—

One peck of fine middlings and half a peck of barley-meal placed in a coarse earthenware pan and baked for one hour, then boiled water is poured in and the whole stirred together until it becomes a crumbly mass—or the baked middlings can be mixed with rice, previously boiled—two meals of this mixture might be given each day, and one meal of grain.

Oats are good for laying hens, but to my mind are best ground; it is not at first a favorite food with poultry, but they soon acquire the taste, and it is even more nourishing than barley, but also more expensive. Oatmeal is considered wonderfully good and fattening diet, and in Ireland is generally used for poultry—that is, when they are kept up for market, the meal is mixed with milk and mashed potatoes. In oats there is as great an amount of starch as in barley, more flesh-forming substance, and more fat-producing matter.

Light Wheat is the grain I prefer for poultry-food; but, alas! it is not easy to procure, though it is cheap as far as price goes. If you have a farmer living near you he may perhaps let you have some as a favor; but, as a rule, farmers keep it for feeding their own poultry, and do not care to sell it at all.

Buckwheat and Hempseed are very good, the latter to be given during moulting, but they are too expensive to be given frequently.

Indian Corn is good and economical food, but too fattening to be used much; as a change, though, it is desirable; its usual cheapness, compared with the price of our home-grown grains, commends it in some places; it should not however, be given whole, but ground into meal and mixed with water or milk.

Linseed is chiefly given to prize fowls and those intended for exhibition; it increases the secretion of oil, and makes their plumage shine and look glossy.

Potatoes steamed and mashed are very nourishing, but rather expensive.

Bullock's Liver boiled and cut up into small pieces may be given with much advantage once or twice a week to birds kept in small inclosures.

Malt is one of the best things for poultry, but not very easy to procure; if, however, you are near a brewery you will not have so much difficulty in obtaining it. It induces early and continued laying; should be given sparingly, either bruised or whole, about two handfuls for every six fowls; it can be mixed with the ordinary food. For chickens also it is desirable, about one handful to every six; if they are fledging it assists them in putting on feathers, and at all times helps their growth.

Milk should be constantly given—that is, where a cow or cows are kept, otherwise perhaps it would be rather an expensive addition to the cost of poultry-keeping; but if the food is wetted with fresh milk, or a little warm milk stirred into the rice or various meals in use, it is astonishing how very much further the food goes, for it gives a satisfying property to it, and is most nourishing, especially for the younger members of your fowl family.

Green foods are all good, and should be given daily: chopped cabbage, clover-heads, turnip-tops, lettuce, turnips, boiled or steamed, form also a good change of diet, and grass fresh cut from lawns, or a handful plucked and thrown into the yard now and then, will be much appreciated. Fowls, as I said before, are by no means fastidious in their tastes; grain, soft, animal, and green foods all come alike to them; worms, maggots, and slugs are also delicacies, but not very often procurable, though French poultry-keepers and others take the trouble to form heaps of earth, manure, dead leaves, and so on, on purpose to generate supplies of worms with which to feed their fowls.

To those who would keep fowls economically, and yet profitably, I say save *all* table and house scraps. If you do not keep a pig you will have plenty for the fowls: crusts of bread, stale pieces, scraps of meat, fish, vegetables, bones broken up, soup bones, after they have been used and their goodness extracted by boiling down for stock, yet contain no small share of nourishment; broken and pounded till small, they are almost necessities for fowls kept in partial confinement.

If you feed fowls on grains and expensive meals you cannot expect a profit from them; but if, on the contrary, you utilize house-scraps—which would otherwise be wasted—and give green food, you will be a considerable gainer; if you have to buy all the food, of course you will find poultry-keeping rather an expensive amusement instead of a paying one.

My poultry family I feed in this fashion—that is, the stock birds—the chickens, of course, have more delicate food, and that more frequently given:—

First meal, given about 7 a. m.—fowls are early risers—is of grain, inferior barley, or wheat-tailings, or meal in a crumbly state.

Second meal, midday, of soft food, pickings, such as bread, sops, meat and fish scraps, with either barley, oats, or Indian meal mixed with it, or else boiled rice, peppered in winter.

Third meal, before going to roost, grain. I vary the food as much as possible, sometimes giving two meals of grain and one of soft food, at other times two meals of soft mixture and one of grain, and at least once a week give chopped liver, well boiled but fresh—not in the horrible putrid state some people suggest. I could not fancy eating a fowl fed on carrion myself, though I know it is frequently done; but the flesh on fowls so fed must, one would naturally think, be gross and rank-tasting.

Water should be plentifully supplied fresh and pure and the pans refilled frequently in summer; in winter all water-pans should be emptied out at night, as, if the water freezes in them they often crack or break.

Lime and mortar rubbish or broken oyster shells should be freely scattered about the yards, also gravel and small stones. Fowls like to pick such things up; besides, it is necessary that they eat some shell-forming material or their eggs will be soft, which is very often the case if such substances are not provided. I do not believe in cooking or grinding *all* the grain foods, and should certainly give wheat-tailings or inferior small barley in its natural state. If the birds could not digest it they would not have been provided by Nature with an elaborate apparatus for softening and grinding it. If we feed entirely on moist food even fowls in confinement, we must weaken the action of the gizzard by not giving it enough work to do. The two extremes of feeding entirely on cooked and moistened food, or entirely on grain or hard food, are both mistakes; vary the food, and allow only one meal of solid grain, which should be given either as the *first* or *last* meal, but do not so completely interfere with Nature's laws, as to weaken an organ which is purposely provided to render the natural food wholesome. By allowing plenty of lime and mortar rubbish in your yards, small stones, and so on, your fowls, even in confinement, will be able to digest a small portion of grain each day. I am well aware that many poultry-fanciers say *cook all food*, but I am certain that too much moistened food is not altogether good. I can only speak from my own experience, and I never found the creatures under my care suffer from eating small whole uncooked grain once a day.

The gizzard is a most powerful grinding-mill, being composed of very thick muscles, and lined with a tough insensible coriaceous membrane. The two largest muscles which form the grinding apparatus are placed opposite each other, face to face, just like two millstones, and they working on each other grind to a pulp the food which is subjected to their action and break it down until it is in a fit state to be acted upon by the gastric juice, which softens the grain. Until, however, it has gone through Nature's grinding-mill the gastric juices have no power upon it to render it solvent. By giving food constantly which does not require the action of this apparatus upon it to render it wholesome we run the risk of injuring it by inaction: this surely stands to reason. In the case of chickens even a little very small grain should be given, that while the gizzard is growing it may have something to act upon, and no grain is so good for this purpose as the tailings of wheat before-mentioned.

It is a bad practice to underfeed poultry, or, in fact, any young stock; but, on the contrary, do not waste food; scatter it for them, and when they cease to run after it stop feeding them, is a fairly good rule to go by. It is said that one full-grown bird will eat half-a-pint of grain each day, because, though it may not positively consume that amount of grain—what with meal-scraps, green stuff, &c.—it consumes food to about that value.

SILVER-PENCILED HAMBURGS.

INCUBATION.

Of artificial incubation I may as well say at once I have had no experience; therefore it is a subject of which I do not presume to write; but I cannot think that it is at all adapted to very small poultry-yards, for it must entail primary outlay, endless trouble and considerable expense. On large farms it may answer, or with persons who are bitten with the poultry mania, love trying everything new that they hear of, and have more money than they know what to do with unless they indulge in some hobby or hobbies to help them in making away with it. The invention of the artificial incubator cannot be considered, however, as a new invention, for as early as 1848 Mr. Cantelo, manager of the Model Poultry Farm at Chiswick, brought out the "Cantelonian Hydro-Incubator," and shortly afterwards Mr. Rouillier invented another—an improvement on the one named. Since then their name has been Legion.

The old natural method of allowing the hen to sit on her eggs and hatch out her small family is the only plan of which I have had practical experience, and as being an entirely natural process I cannot but think it the best, especially for poultry-keepers on a small scale.

There are very many little matters connected with eggs, and hatching them out, which can only be learnt by much practice and long experience of domestic fowls, their manners and habits.

This can only be gained by being constantly with them and carefully watching them through all the various stages of their lives.

It is never very difficult to procure a broody hen. Your Brahma hens will most likely be quite willing to sit, probably more often than you wish them to. Be careful, however, not to put under her at once the eggs which you have selected for your sitting. She should be moved in at night, placed on a sitting of china eggs, and allowed to sit on them for at least two days before you entrust her with real eggs.

Now about the eggs themselves. Probably you have, out of your family of hens, some that are better than the others, either in shape and form, or more handsomely marked, or better layers, or there is something or other about them, some distinguishing point, which leads you to wish to perpetuate their stock. Their eggs should, therefore, be saved; but do not keep eggs certainly beyond a fortnight; the fresher the eggs the better, I believe. Those you set apart for a sitting remove directly they are laid and place them in bran, small end downwards, dating them in ink, and adding the name of the hen. Does this sound absurd? Possibly to people who know little and care less about fowls it may, but those who keep a limited number I venture to say would have their original family of birds named, either by names caused by some distinguishing mark about the bird, or in groups adhering to one initial letter.

When you have collected, say, thirteen eggs, which is quite enough to put under any hen, though people do advise fifteen for a large hen—too many really for a hen, though a turkey would cover them comfortably—thirteen for a large Brahma hen, and eleven for a smaller hen are the number I usually place under the hen, and find them quite enough. If a nest is too full of eggs there is sure to be an accident: some eggs get broken and the nest gets foul and sickly; besides, the hen covers a compact nest of eggs much better, and they all get an equal share of heat.

All the eggs placed under the hen should be marked with their proper dates. Have the eggs as near as possible in date, so that the chicks may hatch out close together. A great advantage of marking the eggs is, that should the hen lay any when first beginning to sit, or should other hens gain access to the nest, the fresh eggs laid can be removed. Mr. James Long, a great authority on poultry, advises that at the end of ten days the eggs should be tested. This should be done in the evening by the light of a lamp, holding the egg betwixt the thumb and forefinger of the right hand in front of the flame, and shading the large end with the base of the left hand, the air-chamber is discovered; this is apparently opaque, the rest of the egg being dark and heavy, the two portions being divided by a clear black line—that is, if the egg is fertile. If, on the other hand, the egg is light and opaque throughout, or, in other words, exactly like a new-laid egg when held before the same light, it is not fertile. This little test is so simple that every one should adopt it, and use the eggs found unfertile, not returning them to the nest. They are just as edible and as wholesome as eggs laid on the same day but not placed under a hen, and can always be used in the kitchen, being quite as good if not better than the so-called "cooking eggs." Sometimes, however, these unfertile eggs are not clear and edible, but rotten; this can generally be detected. If the egg, on being tested in the manner described, is found neither clean nor fertile with the dark line at the top, but without the dark line and dull throughout, especially in the centre, the whole mass within the shell being in a movable state, its condition may be reasonably suspected and it can be thrown away. This state may arise from one or more causes; it is fancied that it arises from the fertilization being incomplete or weak, wanting sufficient strength to break into positive life, but yet enough to affect the rest of the egg, which, as in all cases in which any life has existed, decomposes, and in time engenders gas. Such eggs should be buried, not thrown where they can be picked at by other birds.

It is a good plan to sit two or more hens at the same time; on the tenth day you can test the eggs, and remove from both nests the unfertile ones giving one hen all the other eggs and resitting the other on a fresh lot of eggs. Besides, if two hens sit at once, one hen when they hatch out can take both broods, so you economise your stock of hens. I would never advise, as some people do, that hen No. 2 should be given a fresh set of eggs and have to sit another three weeks, for no hen could sit six weeks without taxing her strength too much; this proceeding I look upon as a downright cruel one.

Short-legged hens are the best for sitting, therefore Brahmas and Dorkings make very good "broody" hens. Three weeks is the usual time it takes for hens' eggs to hatch, but they may either be a day or two before or a day or two after the twenty-one days.

If possible have a sitting-house, or arrange that your sitting hens are kept in a quiet, rather dark place, away from the other birds, else you will have endless trouble; for if kept in the same house in which the other hens lay, they will be constantly interfering with the sitting hens, trying to lay in the same nests, and eggs are sure to be broken in the scuffle. Your sitter may prove a little restless in a fresh place at first, but employ china eggs for her to sit on until she is disposed to sit steadily, and she will soon settle down, you will find, in her new nest, especially if she be really "broody" or "cluck." And here it may be as well, perhaps, to say a few words about "broody" hens. Sometimes they are most tiresome, and very often this strong desire to sit, which is termed *storge*, is so strong that no means you can try will abate it. In such a case I should be tempted, even if I did not want the chickens, to let the poor hen gratify her desire, and do as the French *acouveurs* do. They only provide broods, but do not rear *them*, selling their chickens at twenty-four hours old, and sending them to the *fermiere* who has ordered them packed up warmly in flannel in a small flat basket. Chickens, curiously enough, travel very well at that early age, better even than when they are older, because Nature provides them with nourishment when they first hatch out, and they really need nothing till the next day but to be kept snug and warm. When they reach their destination, which must, of course, be within reasonable distance, they are given at night to a hen who has a brood of chickens of about the same age, who will, as a rule, welcome the addition to her family with pleasure, seeming rather to delight in this mysterious increase to her family. A hen is always very proud of a large brood, and I have often noticed will apparently, in hen language, crow over a less fortunate mother with only a few to take care of.

I once had a hen who had only one chick. She got shut away from her nest by accident, and was kept out so long that the eggs were spoiled all but three, and from these were hatched very weakly chicks. Two died in the act of being liberated from their shells, and the result of the sitting of thirteen eggs was one chick, and that took a considerable amount of cosseting and nursing before it became quite strong. It was most absurd to see the mother, the fuss she made over her one bantling. It was a late sitting, and I had no other chicks ready to enlarge her family. When the chick was a few days old, her favorite mode of carrying it was on her back, and there the little creature sat quite contentedly while the hen marched about. This went on for months, until really the single scion of the house of "Raca" was as strong as his mother. But the affection between the two was too funny. Even when he was a fine handsome cockerel, about to be promoted to reign in the room of his father "Raca" as Raca II., or over another harem, his mother would insist on presenting him with scraps and dainties she had picked up. I never knew a case in which the tie of relationship betwixt hen and chick lasted so long.

To return to the subject of "broody" hens. I certainly wonder why here in America we do not adopt French methods with regard to rearing poultry. We spend days, weeks, in trying to cure a hen of wishing to sit, a perfectly natural inclination, very often starving and really cruelly teasing the poor thing, while all that time she might be fulfilling her end in life, and sitting on a nest full of eggs. She does not cost more while she is sitting, and, indeed, it is far more economical to employ her than to chase the poor wretch off the nests, shut her up, give her physic, or otherwise torment her. You may argue, "Oh, but my hen would lay again soon if I prevented her from sitting!" Pardon me, but the hen certainly would *not* lay under a month, and probably not for six weeks, as she will pine at first and lose flesh from the feverish anxiety of her state, will be some time before she gets in condition again, and very often two or three months will elapse before she will lay; whereas, after sitting, even if her chickens are removed from her or she is only left with one—perhaps you feel inclined to allow her one or two after her trouble of sitting so long—she will begin to lay again sooner than she would were she laboring under the *storge*. If it is very late in the season you might get ducks' eggs and sit your "broody" hen on them. Ducks do better in cold, inclement weather than chickens, and when sold bring in a good price. They cost more to fat, though, as they are such ravenous feeders.

Sitting hens should have a *daily* run. Do not remove them forcibly from their nests, but let the door be open every day at a certain hour for a certain time while you are about. Perhaps for the first day or two you may have to take them gently off their nests and deposit them on the ground outside the door. They will soon, however, learn the habit, and come out when the door is open, eat, drink, have a dust-bath, and return to their nests. That this should be a daily performance is quite necessary to their health and well-being. It is a very old and mistaken notion to fancy that the chicks hatch out better if the hen sits close and never leaves her nest, because it is not so; air, food, exercise, and a roll in the dust are necessary to the hen's health, and the eggs will not come to any harm.

Some people, while hens are off their nests, damp the eggs with lukewarm water. Moisture, they say, is necessary, and the chicks gain strength by the process. This may be correct, and, in very dry weather, perhaps necessary. Myself I never fancied it did much good, though I have tried the experiment; but I consider it is a mistake to meddle too much with nest or eggs; the hen is only made restless and dissatisfied by so doing, and the result is not such a very decidedly good one as to be worth the extra trouble. While the eggs are hatching out do not touch the nests; it is very foolish to fuss the old bird and make her angry, as she treads on the eggs in her fury, and crushes the chicks when they are in the most delicate state of hatching—*i. e.*, when they are half in and half out of the shell, when a heavy tread on the part of the old bird is nearly certain to kill them.

Picking off the shell to help the imprisoned chick is always a more or less hazardous proceeding, and should never be had recourse to unless the egg has been what is termed "billed" for a long time, in which case the chick is probably a weakly one and may need a little help, which must be given with the greatest caution, in order that the tender membranes of the skin shall not be lacerated. A little help should be given at a time, every two or three hours; but if any blood is perceived stop at once, as it is a proof that the chick is not quite ready to be liberated. If, on the contrary, the minute bloodvessels which are spread all over the interior of the shell are bloodless, then you may be sure the chick is in some way stuck to the shell by its feathers, or is too weakly to get out of its prison-house.

The old egg shells should be removed from under the hen, but do not take away her chicks from her one by one as they hatch out, as is very often advised, for it only makes her very uneasy, and the natural warmth of her body is far better for them at that stage than artificial heat.

Should only a few chicks have been hatched out of the sitting, and the other remaining eggs show no signs of life when examined, no sounds of the little birds inside, then the water test should be tried. Get a basin of warm water, not really hot, and put those eggs about which you do not feel certain into it. If they contain the chicks they will float on top, if they move or dance the chicks are alive, but if they float without movement the inmates will most likely be dead. If they (the eggs) are rotten they will sink to the bottom. Put the floating ones back under the hen, and if, on carefully breaking the others, you find the test is correct (one puncture will be sufficient to tell you this), bury them at once.

Chickens should never be set free from their shells in a hurry, because it is necessary for their well-being that they should have taken in all the yolk, for that serves them for food for twenty-four hours after they see the light, so no apprehension need be felt if they do not eat during that period, if they seem quite strong, gain their feet, and their little downy plumage spreads out and dries properly. Their best place is under the hen for the time named, then they may be fed in the manner described under the head of "Management of Chickens."

MANAGEMENT OF CHICKENS.

Chickens will, as I have already said, do without food for the first day and night; but as soon as they begin to feed they should be very well fed, and constantly. We all know the old saying, that "Children and chicken are always picking." At first their food should be crumbs of bread, sometimes dry, sometimes soaked in milk, and the yolk of hard-boiled eggs crumbled up and mixed with bread-crumbs. This is quite enough for the first week. Afterwards small grain may be given, chicken wheat, or tailings of wheat, groats, canary-seed, a very little hempseed, bits of underdone meat minced small, a little finely-chopped green food, macaroni boiled in milk and cut into small bits, and so on. They should be fed very often, but only given a little at a time. I feed mine every two hours for the first three weeks or so, taking care that they only have just as much as they can eat at a time, so that the food is not wasted. Hempseed must be given with caution; but if the weather is cold and damp it is very good for warming the chicks, and they are very fond of it. Soft food mixed dry should be given them after the first week, macaroni, barley-meal, or middlings. This mixture should be made with milk, or, if no milk is given, then scalded water, but on no account should any food for chickens be mixed with water which has not been *boiled or scalded*. The food should not be mixed in a wet, sloppy mass, but of such a consistency that when thrown on the ground it will crumble readily.

The old hen should be supplied with grain (wheat), some of the meal, or any other food suitable for her when her little ones are fed, but not oats. All water which is given to the chickens should be boiled first, or else it is very apt to give them diarrhœa. A very good drinking-pan can be made for the small birds by inverting an ordinary flowerpot in its saucer, and filling the latter with water. In this they cannot drown themselves, as they might in a deeper pan or ordinary drinking-trough. Many people give skim-milk instead of water at first.

All the time chickens are growing they should be well fed. It is the very greatest mistake to stint any young stock; and chickens, if you wish to bring them on quickly for market, must be well and generously fed at all ages, not neglected when three-parts grown, as is too often done. They should be constantly supplied with fresh water.

It is certainly best to confine the hen under a coop for the first month or so. If she is allowed her liberty she will wander about with her brood in search of insects, and so may expose her family to the attacks of hawks, weasels, or other vermin. And, besides this, though you wish to feed your hen well while with her brood, it would be rather foolish to allow her to satisfy her appetite on the dainties prepared for them, which she naturally will do unless you give them their meal where she cannot reach it, but giving her under her coop at the same time coarser food. Economy points out that delicate and expensive food, such as groats, boiled eggs, and crumbs of bread, should be reserved for the chicks, while the hen has wheat or ordinary food. I should not feel inclined to give her oats or barley unbruised for this reason: she will, of course, call her little ones joyfully to her to partake of the food given her, and they might choke themselves with large whole grain, such as oats or barley. Rice will not hurt them (boiled, of course), nor wheat, which is a much smaller grain, especially in tailings, than the other cereals mentioned, and cannot injure her little family, even if she does give them a grain or two.

For the first week or two the coop should be placed in a warm sheltered spot, but taken into a safe place at night. As the chicks gain strength it should be moved on to a grass plat. The ordinary-shaped coop, with a sloping roof and barred front, is as good a one as any, only I should advise handles, strong wooden ones, being fixed to each side to facilitate movement. Boarded bottomed coops are not desirable—it is far better to place a bottomless coop on the ground—else you might have small wheels to your coops to push them along when changing their place. In a case of emergency, or if the expense of a coop cannot be incurred, an old cask or beer-barrel makes a very fair coop. Knock out one end and put laths across, leaving one to draw in or out, and take out the staves which rest on the ground. The barrel should be propped on each side to prevent its moving, and a tarpaulin must be provided to throw over it at night to prevent any rain soaking in the knocked-out end, and will serve as a cover for the opening, which must be closed, for fear of cats, foxes, rats, and such creatures. Holes for ventilation must be drilled in this cover. I have reared many a healthy small brood in a barrel in this way. It is easily rolled, too, into a fresh place, and if you have not coops enough, and do not know where to stow your small families, barrels or boxes must be turned to account.

The chicks when about a week old should be allowed a little liberty. The old hen might be turned out with them for an hour or so during the warm part of the day, only she must be watched in order that she does not lead them into mischief. About this time, too, their food should be changed; less soaked food and more small grain be given instead—grits, boiled barley, and other articles of diet before advised.

MANAGEMENT OF CHICKENS.

Chickens should never be let out too early in the morning even when they are three weeks or a month old, as it is certainly bad for them to be about while the dew is on the grass.

The coops should be constantly changed about from place to place, but never allowed to stand on wet, moist ground. One of the great secrets in rearing chickens is *always to keep them dry*. If they are allowed to be out in the wet, or kept on damp ground, they will soon become delicate. "Gapes," that fatal malady will attack them, or diarrhœa, or some other ailment, and they will soon die off.

When they begin to feather the very greatest care should be taken of them, as this is a very critical period. Hempseed and bread soaked should be given, and iron in their water. At six months they should be in full plumage, and in seven or eight months the pullets, if they have been well fed up to this time, will commence laying. "Tailings" (wheat) are really the best grain food for chickens up to four months. After they first begin to eat grain many people advise barley, but if you can get wheat—which is not, however, always easy to procure—I infinitely prefer it. If you must give barley, then let it be bruised.

I am no friend to keeping chicks indoors, as some people advise, for I am convinced it makes them weakly. Find a sheltered corner for the coop, and move them into it even in cold weather, only put the coop under shelter at night. Confining them indoors, even in a barn or a stable, appears to produce cramp and weakness of the legs, which when turned out is not the case, for the best and surest preventive for cramp and leg-weakness is to let the birds so affected have their liberty in the air, where they can get the exercise they really require.

With regard to the time for chickens to be hatched out, I rear young chickens most months in the year, but then my fowl-house is in a sheltered place and on good dry soil. If you sit late in the autumn, say in October or November, it is an advantage—that is, if you are in a fairly-sheltered, warm, dry spot—for chickens hatched in December and January bring in a handsome profit in the shape of "spring chickens." There is, of course, a good deal of risk, and immense care must be taken of the young birds during the cold weather, but if the situation is good it is well worth a trial.

By the end of June, or early in July, pullets hatched in December should, if they have been really well fed, be commencing to lay. Your early chickens will not, perhaps, be as strong as those hatched two months after, say in February and March. This is one reason why they should be reared for table. In any case I should not breed from birds hatched in the coldest winter months; but in the case of pullets, if I did not kill them all off as "spring chicks," fatten and kill them when they had finished laying, and before they began to moult; for birds hatched, say in March or April, would be really much stronger, and "selected" ones for keeping out of such broods be more to be depended on to supply the place of some of the old stock if you mean to kill any of them off.

If you wish to fatten "spring chickens" quickly for market, when they are about two months old confine them in coops and feed chiefly with moist food. In my opinion a fowl allowed its liberty has a better flavor than one confined and fed up in a coop, but it certainly does not put on flesh so quickly nor yet get so thoroughly plump and tempting-looking when trussed ready for market, therefore I should advise that those chickens fattened for sale should be kept in coops and fed up, while those for home use should be allowed their liberty until they were really wanted by the cook.

With regard to foxes, rats, and such vermin, your best safeguard against them is to house all your stock at night, and see yourself that their numbers are all right.

Rats must be waged war against. They are the greatest enemies to young ducklings, also chickens. Keep steel traps prepared, putting them down when the fowls are shut up for the night in the runs outside, and baiting them with cheese or bits of meat, only drop a little oil of rodium on the bait. In time, if you will persevere, you will either frighten them away or else catch them; but you must of course, keep your traps out of the way of the fowls themselves. Boiling coal tar poured down the holes, and followed by a deluge of water, is said to be very effectual in making rats desert a yard. I am averse to poison, because, if it is used in a fowl-house or yard, it is next to impossible to prevent an accident sooner or later. Ferreting every now and then will do good, unless your houses are adjoining barns or extensive outhouses, in which case you are more likely to lose your ferrets than destroy the vermin. The holes should be carefully stopped with a mixture of ground glass, bits of glass broken up, and ordinary plaster. Rats will not often attack glass mixed in this way. If you do use poison you must nail it up somewhere out of reach of the birds. This you can do by getting a small bit of meat, soaking it with the poison, and nailing it on to a bit of wood, nailing that again to the wall. Myself I should be afraid of the rat, in his efforts to get off the meat, dropping little bits of it on the floor, when of course the fowls would be the sufferers.

Cats are enemies also. Dogs one has not much reason to fear, but in the vicinity of a town cats of all ages and sizes will sooner or later visit you, and if there is one delicacy they prefer to another it is a young chick or duckling. They are so cunning too, it is hard to catch them.

Traps are not much good. Poisoned fish put down near where you fancy they get into the run is the only thing; but of course it must be taken away without fail before you let your fowls out of the house, and it should also be nailed to a piece of wood, which might be smeared with oil of valerian—of which some cats are so fond—to make it even more attractive. I never lost any chicks by cats, I am bound to say, and therefore should be loath to set poison down for them. I dread poison too, as I have already said, in a fowl-yard. One cat I had who took a fancy to a young duckling, but was discovered before she had eaten it, so poor ducky was tied to her neck in such a position that she could not get rid of it, and this effectually cured her of killing ducklings or chicks. A good hungry half-starved town cat, however, one could not cure by such means; it would be a case of "first catch your cat." But still cats I look on in a light of friends, unless I suffered too severely from their attacks I should not like to demolish them by such a cruel method as poison.

FATTENING.

In feeding fowls for table, or rather for market—for I should never coop chickens to fat merely for home use, as I have before said—much depends on circumstances.

Spring chickens should be penned for fattening directly the hen shows a desire to leave them, when they are, say, five weeks old. They will not then have lost their first plump condition, and will soon, if well fed, increase rapidly in weight. They are not required to be very large; indeed, if fatted too long buyers would fancy they were not really "spring chickens," which frequently make their appearance at table not much larger than blackbirds, and are then considered all the greater delicacy.

If your chickens were hatched out in December, early in February you can put them up to fat. Their coops or cages should be placed in a warm dark sheltered place. There are a variety of different coops or pens recommended by different authorities on poultry to fat chickens. I do not remember to have seen one, however, which is, to my mind so suited to the purpose as this of which I give a description.

As far as the general conformation of the coop goes, it is made on the same plan as many others; but the adaptable shelf which is its chief feature is entirely my own idea, and if adopted would, I feel sure, give general satisfaction.

The coop itself is made of boards, sides, back, and ends of front; centre of front is barred, with the two middle bars movable. It stands on legs between two and three feet in height, the roof sloped sufficient to allow the rain to run easily off. To hold four chickens at the same time the coop should be about five feet in length, four in breadth, and three in height—that is, above the legs on which it stands. If the birds are kept in separate divisions then a little more length will have to be allowed for the partitions. This will give ample room for the birds without uncomfortably cramping them.

The bars in front of the coop should be wide enough apart to allow the birds to get their heads through easily to get at their food, which should be given them on a shelf or board. The shelf, when not in use, being fixed on hinges, would fold down in front of the coop. This is a much better plan than having a trough for food *fixed* outside, as so many coops have, the objection to it being that the food soon gets sour—I mean what is left after the birds have fed—sticking to the sides of the trough, which, if it is a box-like fixture, it is next to impossible to clean properly.

The shelf should have an upright lath nailed to it to prevent the chickens pushing the food beyond their reach in efforts to get at it, but the ledge should not be so deep as to interfere with the shelf closing against the front of the coop. Between this ledge and the coop should fit a zinc trough, the width of the division, for water.

When food is put down this water-trough should be slipped out, to be replaced when the meal is over. Two small wooden supports would prop up this miniature table; on the same plan an extra shelf is made to enlarge an ordinary table. A further use of this flat board would be to close up the front of the coop at night. It should not close entirely the barred space, room being left at the top for ventilation. Water not being required at night, the zinc trough should be removed to allow of the shelf being closed, while the wooden buttons would keep it firmly in its place, the small holes at the side of the coop supplying the extra ventilation necessary.

With regard to sanitary arrangements, before the birds are put into the coop it should be thoroughly washed with a mixture of lime and size, to destroy all vermin. This dries quite hard and does not rub off. If a *white* wash is objected to in a feeding-pen, then it could be darkened by the admixture of color, only see that there is no lead in the color mixed with the wash to procure the darker shade. The partially-boarded front will prevent the coop being too light.

The floor should be first of all of flat bars placed length-wise—fixtures these—and over them should slip in, from the back of the coop, a movable board, which should be drawn out every day and thoroughly scraped and sanded, but not washed, because if not thoroughly dry when put in the birds would get a chill, and very likely suffer from diarrhœa in consequence.

If after a meal there is any food remaining, let down the shelf and brush it off, giving it to the other fowls in order not to waste it. Food should never be allowed to remain in the sight of fattening fowls, or they will lose appetite. If they are only fed at stated times, and when they have eaten as much as they require the board is carefully cleaned and the water-trough replaced in the niche, the birds will feed again, when the time comes round for their food, with eagerness, which will not be the case if the food is left there for them to peck at.

I have had plenty of experience with fowls, having reared them for show, for eggs, and for table, and have therefore no hesitation in recommending my "adaptable shelf," as I feel certain it is an addition of the greatest use to an ordinary feeding-coop. It adds very little to the expense, is so simple that any carpenter could easily make it from a plain drawing, avoids waste of food, and insures cleanliness. As soft food is mostly used in fatting chickens, it is all the more necessary that none of it should be allowed to remain after the meal to turn sour, disagree with the birds, and take away their appetite. In a trough it is hardly possible to prevent a little lodging in the corners and sides, as if the trough is a fixture it cannot be removed to be washed; on a shelf remains of food need never be left, as the application of a hard brush for a few minutes would remove every particle, a little sand being afterwards sprinkled lightly over the board to render it perfectly sweet before the water-trough is slipped in.

Water should be constantly changed, and *boiled* water should be always used instead of that just pumped or drawn from a well or spring, as this will prevent the chickens getting diarrhœa.

You should have some plan of darkening your pens, either by letting down a tarpaulin over the top or having sliding boards to run in and out, so that the light can be regulated at will.

Some people keep their chickens separately, having their pens divided. I do not think this is really necessary if you choose chickens of the same brood to fat together. Four are enough to fat at a time; but never allow your coop, if you have only one, to remain empty; as you kill off one lot of chickens you should have another batch ready to put in. Cramming I am entirely averse to. It is a needlessly cruel and disgusting custom, though very frequently practiced, especially in France.

Now comes the grand question of food. It should always be pultaceous; the birds cannot pick up pebbles and little stones when shut up, so cannot, digest grain of any sort. Feed them on bread and milk, oatmeal and milk, rice well boiled with a little pepper mixed with it, barley-meal, Indian corn meal, potatoes steamed and mixed with barley-meal, chopped green food, &c. Very many breeders give a large amount of suet mixed with the food, but unless people are fond of greasy fat on their poultry, which to me is an abomination, I should not advise it, as it makes the flesh so gross. *Vary the diet as much as possible,* and never give it in a sloppy state, but crumbly. Three weeks or a month at the outside is enough to keep fowls up for fatting: if kept longer the confinement begins to tell on them. Some people mix treacle or sugar with their food. Saccharine matter is no doubt conducive to fat, and oatmeal, or Indian corn ground into meal and mixed with treacle until it is in a crumbling state, is a food all chickens are fond of, but should only be given to those you wish to feed; it would not do for those pullets you wish to bring on to lay quickly, as it would develop interior fat, which is always fatal to constant laying.

Guard against waste of food. Only experience will cause you to know how much to supply at once; and until you learn this, directly you see the chickens begin to pick daintily at their food remove it, give to the other fowls then what is left, but on no account allow it to stay in the trough for the fatting chickens to eat, when, as the old women say, "they've a mind to." If they do not constantly see food before them they will eat it far more readily when it is given. This is only common-sense treatment, and, believe me, in dealing with fowls you must often draw largely on this very desirable commodity.

Four meals a day should be the allowance for penned-up chickens, letting them eat, each time you feed, as much as they will with appetite. At night they will roost on the board. Some people put down clean straw, but if you close up the pen so that the birds are not cold it is not really necessary, and it only harbours insects. Perches you might have if you've room in your pen—sufficient height I mean. Before the birds are put in have the coop well cleaned, white-washed, and sprinkled with carbolic acid. This should be done two or three times during the time the chicks are fattening.

Fowls should of course be killed in the most merciful way. It makes one shudder to read of the manner in which the poor things are sometimes tortured, allowed to bleed slowly to death, pins run into their brains, and horrors too dreadful to name. Poultry dealers generally kill them in the quickest manner by breaking their necks, and so quickly do they perform their work that one man will often kill and pick a dozen or more in an hour. One of the easiest ways of killing is to hit the bird a sharp blow on the back of the head with a heavy blunt stick; death is almost instantaneous. Then pluck at once while the bird is warm, as the process can then be accomplished much more rapidly than if the bird is allowed to hang until cold. When all the feathers are off the fowl will still be warm. It should then be carefully singed, floured, and trussed, and placed between two boards with a weight, on the topmost; not too heavy a weight, of course, to spoil its shape, but just enough to keep the breast down and in good shape.

Capons of course fetch much better prices, and their flesh remains tender up to the age of two years, whereas a cock at that age is only eatable in a stew, or pie. Chickens converted into capons increase in size to a wonderful extent; the birds will in a year be nearly treble the size it would have been if left alone, and double the market value.

In conclusion I may observe that I can most sincerely, from my own practical experience, advise all ladies, as well as gentlemen, who have a little room to spare in their back gardens, to set up poultry-keeping on a small scale. Many more people keep fowls now than used to years ago, I know, but still not half people enough. Many who have room to spare for a family of fowls let that room remain unoccupied, either from a mistaken idea that poultry-keeping is too expensive or will entail too much trouble on them. With regard to the latter idea, it is, no doubt, a partially true one. Fowls do cause trouble, and if they are to be made to pay their way cannot fail to do so. But whatever trouble they cause they are worth it, and no undertaking or pursuit that I ever heard of flourished without some amount of trouble. In return they give fresh eggs—that you are sure of, and can offer a guest without any inward misgivings—plump chickens, a little pocket-money, and a great deal of interest.

PART II.

DISEASES OF POULTRY.
ARTIFICIAL INCUBATION.
Miscellaneous Poultry Receipts.

DISEASES OF POULTRY.

There are very many diseases which attack domestic fowls, the chief being

    Roup, or Glanders.
    Cholera.
    Diarrhœa.
    Indigestion.
    Leg-weakness.
    Gapes.
    Gout.
    Asthma.
    Megrims.
    Apoplexy.
    Bad feathering.
    Moulting.

While among the less frequent ailments may be classed the following:

    Crop-bound.
    Egg-bound.
    Mange.
    Oon-lush, or soft eggs.
    Dropsy.

Torpid-gizzard.
Pip.
Influenza.
Inflammation of mucous membrane.
Paralysis.
Phthisis.
Rheumatism.
Cramp.

And besides all these ills to which fowl-flesh is heir must be named the various accidents to which birds are liable, and external ailments; for example:—

Bruises.
Fractures.
Obstruction of rump-gland.
Corns.
Ulcers.
Elephantiasis.
Bumble-foot.
White comb.
Vermin.

A sufficiently long catalogue certainly; but happily, though this list of diseases, &c., may seem formidable, if fowls are really well attended to and have their general health decently cared for they do not often develop disease, and there is always the one best remedy for a sickly fowl, which is, to kill it. In writing thus, I trust I shall not be thought cruel. In reality it is far more human to kill a bird at once that is severely injured or develops an incurable disease. In the former case the poor thing is put quickly out of its misery and suffering; in the latter the other fowls run the risk of catching the disease, and as many fowl diseases are infectious or contagious, this is a matter for consideration.

Myself I never keep fowls—that is, ordinary birds—if they

seem sickly, but have them killed at once. If they are valuable prize or stock birds, then I try physic. Unless in the case of really valuable birds, it is very poor economy to keep unhealthy ones.

Chickens should never be reared from weakly parent birds for it only causes disappointment, degenerates the breeds, and lessens the gain on poultry-keeping. Take care, therefore, to have your stock healthy, and raise chickens only from really strong well-formed birds, free even from accidental blemishes. Feed judiciously both young and old birds. Keep your houses and yards perfectly clean. Sometimes change the run on to grass if possible, if fowls are not at liberty, or turn them into a field or meadow for a few hours every now and then. Let them have plenty of gravel and lime, dust baths, and perfectly fresh water every day, in the summer twice. Do not have too many chickens at once, or you will not be able to attend to them properly. Always keep your stock of old birds up to a certain number, as it simplifies matters, and you can reckon better the amount of food they require and what they really cost you. Never keep birds after they have passed their prime, because if you do the profit side of your poultry book will assuredly be an unsatisfactory study.

If you attend to these various necessary precautions, which I give as the result of my own personal experience, you will not be very likely to require to doctor your birds. A knowledge, however, of the symptoms of the various diseases and the best accepted method of treating them is necessary, even should you not require to practice that knowledge, therefore I have enumerated the different maladies, and as far as possible their proper treatment.

ROUP.

Roup is contagious. Symptoms: Hoarseness and catching the breath. If very bad, a discharge from eyes and nostrils. Treatment: Give a tablespoonful of castor-oil on perceiving the very first appearance of the disease; take the scale off the tongue with your nail; give soft food only and doses of castor-oil. Wash each morning and evening the eyes, nostrils, and inside of the mouth with vinegar, and give sulphate of iron either in pills mixed with cayenne pepper and butter or else put the powder in the water they drink. Keep the bird or birds so affected away from the other fowls, or you will have the disease communicated to your whole stock.

The very first symptoms are swollen eyes, then a discharge from the nostrils, first clear, but soon changing to thick, offensive matter. Water also is discharged from the mouth. Directly you preceive any such symptoms remove the bird from the others into a warm but well-ventilated place, and then treat as I have advised. Ducks and geese also suffer from roup, and frequently die off very rapidly from it.

Mr. James Long, in his excellent and valuable book on poultry, says that a "solution of chlorinated soda or carbolic acid, the latter in the proportion of 1 to 60," should be used to clear the mouth, eyes, and nostrils of birds suffering from roup. "In bad cases," he writes, "I have found that an injection of the soda through the slit in the palate has proved most advantageous, this being the only solution I know that will really dissolve the pus. The injector must suitable may be purchased at any chemist's, where it is sold under the form of 'a cure for toothache.' It is a small glass tube bent at the small end, the large end being fitted with an india-rubber bag which contains the nostrum. This should be squeezed out and the soda drawn in as often as required by the suction of the empty bag." The following receipt for roup is a very good one:—"Sweet oil 1 oz., camphor 1 drachm, carbolic acid 12 drops. Pulverize the camphor in a mortar with a little ether. Applying with the glass tube twice daily, injecting the mixture into the nostrils, mouth, and through the roof. During the illness cod-liver oil capsules will be found strengthening."

CHICKEN CHOLERA.

This disease is much to be dreaded by all who keep poultry. The causes of it are uncertain, although the crowding of fowls in a small space, unwholesome or irregular food and keeping them on swampy land is supposed to originate the epidemic.

Symptoms: The fowl droops, is weak and much prostrated, and the feathers present a rumpled appearance. Diarrhœa at first very light, increasing in severity and of a green color. Food is refused, and if the disease is not promptly arrested, it spreads among the flock and death ensues on all sides.

Remedy: Fresh green food daily; powdered alum mixed with meal about one teaspoonful to a quart of meal. If an improvement is not noticeable in a short time give each fowl a half teaspoonful of castor-oil and ten drops of laudanum.

After the disease has been arrested spade up the grounds, disinfect with carbolic acid, whitewash and let lime lay around the yards and runs, prevent the fowls from drinking anything but pure clean water and feed on green and wholesome food.

If the bird shows no signs of recovery after the daily doses of castor-oil, together with special feeding and frequent washing of the head and face with injection of the soda, have been perseveringly tried for some time, then it is far better to kill it, even if a valuable bird, as it will only linger on and die a miserable death; besides, the longer you keep it the greater risk you run of your other birds catching it, as they may do this by accident, however careful you may be in separating the diseased bird.

DIARRHŒA

Is generally caused by too much soft food. The diet should be changed directly the symptoms of this malady are perceived. Chalk given in the water, and dry food, such as barley, rice boiled, but the grains distinct, neither watery nor sodden, with cayenne pepper mixed with it, will generally check the disease. If, however, such treatment does not succeed, then try this; five grains of chalk, two grains of cayenne pepper, and five grains of powder rhubarb, made into butter pills. If, however, this even does not succeed in checking the diarrhœa, then give half a grain of opium and half a grain of ipecacuanha every six hours. Keep the bird in a separate place, and warm, as a chill would probably finish it, or being in a damp, cold place.

Chlorodyne I have myself tried with great success, but only when the malady had reached a severe stage. Six drops in a teaspoonful of warm water would be the dose for a full-grown bird, from two to four for a chick according to age. There can be no harm in trying the remedy, because if it does not cure it generally induces sleep, and relieves the suffering bird in that way.

With diarrhœa the best thing is to check it at once; if allowed to go on it often proves fatal, but by prompt and judicious early treatment it can generally be fought against. The bird will, however, even when cured, have to be taken care of with regard to food for some time, and should have some sort of tonic given it, in the water it drinks.

INDIGESTION.

Indigestion is an ailing to which highly-fed fowls kept in confinement are very often subject. It is caused from failure of macerating power in the gizzard. A teaspoonful of castor-oil may be given, if the disease is suspected, every other day, and the bird fed on different food. Lime-water has been tried with advantage, and old mortar pounded and mixed with meal may be given instead of their ordinary diet. Sulphur and cayenne pepper in the ratio of, six parts of sulphur and one-sixth of the pepper, and mixed with barley-meal. The birds should also be allowed more exercise, a run on the grass or in a walk where can pick up gravel and grit, for it is very often the difficulty of procuring such rubbish which makes fowls kept in confinement suffer, the gizzard being weakened in its action by want of its proper natural food.

LEG-WEAKNESS.

The bird's legs should be bathed frequently in cold water, and tied up in wet rags, the rags kept wet. Iron tonic should be given with its food, and three teaspoonfuls of castor-oil at roost-time twice or three times a week, alternately with oatmeal pills mixed with port wine. This prescription was given me by a great fowl-fancier.

Ordinary fowls do not suffer the leg-weakness as a rule; but fat, heavy birds that have been fed up for shows are very liable to it, and young growing cockerels of the Brahma and Cochin species.

GAPES.

All young chickens are subject to this disease, and it very frequently proves fatal. Old birds are very rarely, if ever, attacked by it. It is caused by small red worms in the throat. Some people maintain that gapes may be called bronchitis, and proceed from inflammation of the trachea or windpipe arising from a derangement of the digestive organs; but whether it be called gapes or bronchitis the presence of small worms in the throat is what causes the suffocating feelings from which the birds suffer, and will die if not relieved. They can be removed by passing a tail-feather, stripped of feathers to within about an inch of the end, down the windpipe. This feather should be turned round quickly and drawn out again, when the little worms will be found sticking to it. Some people smoke the chickens, but, to my idea, this only tortures the poor little things, and the introduction of the feather, if persevered in frequently during the day, very often effects a cure. I have cured chickens myself by this process, giving them *boiled* water with a little iron tonic in it, and cod-liver oil capsules, feeding them with hard-boiled egg and well boiled rice during the time they were under treatment, and one teaspoonful of castor-oil—this for very young chicks; two for those older, about six weeks—every other night.

The red worm found in the throat is now known to be the *syngamus trachealis*. I have found the following preventative to be a very good one; mix together the following compound, lard 1 oz. flower of sulphur ¼ oz. coal-oil ½ oz. Anoint the head of all chicks hatched with this ointment as they are taken from the nest.

In some localities it is a complete epidemic, in others it does not appear at all which would lead one to suppose that the chicken picked up the worm, or it was found in the water; this, too, accounts for the old idea that chickens should never be allowed out in the early morning while the dew was on the grass, for fear they should get the gapes. Many people, too, never give their chickens water at all to drink, but *milk*, and this, though rather expensive, is without doubt a very good plan.

GOUT.

Gout generally occurs in old birds, and is rarely cured, though I did once patch up an old bird so troubled. The best plan to pursue really, is to boil the bird down for soup, for of course this malady does not prevent the birds being good for food. Sulphur in pills has been recommended as a remedy, but with small success I should fancy. I was not successful in curing or even relieving the bird I tried it on.

ASTHMA.

Asthma is a complaint to which old birds are subject, though some-times it effects young birds; but this is, I fancy where there is an hereditary predisposition to it. It is very rarely curable, the best thing being to distroy the bird so afflicted. It arrises from a thickening of the bronchial tubes from previous inflamation, and is frequently accompanied by a positive alteration in the structure of the cellular tissue of a portion of the lungs. It will be easily discovered when a fowl has asthma. The breathing is very difficult, and a wheezy, rattling sound can be detected with each inspiration.

MEGRIMS.

Megrims is a giddiness which causes the bird so affected with it to stagger and fall about when walking, even to run round and round in a circle. A Poland hen I had used to run in a small circle in this way. She was incurable, and had eventually be killed, though, as she was a well-bred bird, I tried different methods of treatment, but without success. This malady is occasioned, it is said, by too good feeding and too little exercise. Castor-oil in tablespoonful doses is the best remedy, and if taken in time might very much relieve the fowl for the time, but I question any permanent cure existing. Careful diet and warmth are essential.

APOPLEXY.

Apoplexy also results from too high feeding and too little exercise. Fowls kept in close confinement frequently drop down suddenly and die of this disease. A large dose of castor-oil should be given as soon as possible, or the bird should be bled from the axillery vein, but this is not an opperation which can be performed by every one, and should certainly not be attempted by a person ignorant of the proper method of performing it. Usually apoplexy is rapid in its effects, and the bird drops and dies then and there when it is taken with the fit; and it is just as well this is the case, for a recovery is extremely doubtful and a bird once attacked by the malady is always subject to a return of it.

BAD FEATHERING.

Bad feathering in chickens should be treated with more generous diet: hempseed, bread soaked in ale, and iron given in the water. Some sorts of fowls are, when young, much more backward then others in getting their first feathers; Cochins and Brahmas in particular suffer from this cause, and may often be seen running about when three, four, or even six months old with hardly a feather on them, presenting a dreadfully bare and miserable appearance. This is more particularly the case when they are hatched late in the year and have the cold winter months before them. Warmth and good feeding is the best thing to be done for them, nourishing food often stimulating the growth of the feathers.

MOULTING.

Though this is a perfect natural process, it is very often attended with very serious results to fowls if they are not looked after during the time they are going through it. It is essentially a wasting period, and the birds should be fed generously and given extra stimulating food while they are losing and renewing their feathers. Unless very sickly they do not lose their appetites, which should be encouraged. In their natural state, or when allowed comparative liberty, there is little or no danger, but when kept in confinement moulting often becomes a critical time with them, and as they increase in age so does their moulting become more weakening. They want, as I said before, more stimulating food, and should have coarse black pepper mixed with their ordinary food, bread soaked in ale. Old nails and bits of iron should be put in their water-pans. I give mine sometimes, during their moulting, Douglas's Mixture, which was recommended in the *Field*. It is made by dissolving half a pound of sulphate of iron and one ounce of sulphuric acid together, and adding two gallons of spring water, and should be given to the birds in the proportion of one teaspoonful to a pint of water instead of their usual drinking-water. It is an excellent mixture, and good for roup, leg-weakness, and other poultry complaints, as it gives the fowls stamina, and should be given not only when actual disease is present, but also in the spring and fall of the year, when they appear drooping or lose their appetites.

CROP-BOUND.

In a case of this sort, which is very easily detected by the abnormal swelling of the crop, the only remedy is to make a small incision in it with a sharp penknife, remove the mass of undigested food which causes the evil, sew up the cut with fine silk, give castor-oil, and feed for a few days on soft food only.

EGG-BOUND.

Egg-Bound is generally caused by the white coagulating in the oviduct and obstructing the passage. When this is the case, the yolks of the eggs, as they become mature and fall into their proper place, cannot pass by reason of the stoppage, which they of course enlarge and increase, until the accumulation becomes putrid, mortification supervenes, and the bird dies. Unfortunately, for this complaint there would seem to be no remedy, for its presence cannot be detected until the mischief is done and it is too late to treat the bird with any benefit. Sometimes, though, the stoppage is merely caused by general inflmation of the egg-organs, in which case calomel and tartar emetic have been known to do good, in the proportion of one grain of calomel to one-twelfth grain of tartar emetic. When a hen lays soft eggs, if she has plenty of shell-making material supplied for her, inflammation may be suspected, and she should be carefully watched and treated in time.

SOFT EGGS.

Soft eggs are often laid by hens when they have not a sufficient quantity of lime and mortar rubbish supplied to them, without which they cannot produce enough calcareous matter to form shells for their eggs. If, however, this is caused by a defection in the ovarium of the hen—and this is more often the case than people fancy—then the following treatment should be followed: Feed on oats instead of barley or other grain, and give the hen a teaspoouful of prepared chalk every other morning for about three weeks or a month. The chalk can be given dissolved in the drinking-water. Do not feed on oats each meal, but give two out of the three meals of coarse meal, and mix with it two teaspoonfuls of old mortar pounded up fine. By the end of a month's treatment your hen will probably be nearly cured; afterwards supply her well with lime broken bones, mortar rubbish, as, indeed, should always be done.

MANGE.

Mange is a malady which attacks fowls in the same way it does dogs or cattle, but is not a very usual complaint; it arises from uncleanliness and too little food, which induces debility; Fowls in good health have smooth, glossy, and fine plumage; when the feathers, instead of having this appearance, are ruffled and stare, in the same manner a horse's coat does, then they are nearly sure to have some malady, very frequently some skin disease, unless they are moulting, and then their plumage is always more or less disordered and unkept-looking. A better class of food, pure water, and the mixture before mentioned put into it, will in such cases give relief, and most probably effect a cure. Fowls with this complaint must be kept apart from the others.

DROPSY.

Dropsy is a rare disease, and fowls afflicted with it had much better be killed at once and put out of their misery, as it is an almost hopeless complaint and one very seldom cured, though the bird may linger on with it for a long while, and suffer very much until death puts and end to its troubles.

TORPID GIZZARD.

Torpid gizzard is caused by feeding too much on soft food, thereby destroying the action of the gizzard, which, having no work to do, becomes inactive and torpid. Change the diet, allow the bird more exercise, and dose with castor-oil in moderate doses at roost-time. I have before expressed my opinion that it is a mistake to feed on too soft food, giving all grain ground to meal, and so on. In their natural state the birds would never have soft food, except worms and green food, and though in confinement it is necessary to give a good deal of the food mixed up and cooked, still grain in its natural state should never be entirely withheld, or your birds will certainly suffer from torpid gizzards and other ills of a like nature.

PIP.

Pip is a very fatal disease; it consists of inflammation in the tongue and throat, and is caused, it is thought, by irritation of the mucous membrane of the alimentary canal. Chickens are mostly attacked by it. Some people scrape the scales roughly off the tongue; if this plan is adopted it should be done very gently with the nail; but *borax* is the safest remedy; it should be dissolved in tincture of myrrh and water, and applied with a paint-brush several times a day. The bird should be kept in a warm place, fed on bread and milk with green food, given plenty of pure water, and dosed with castor-oil in doses according to its age and strength. Pip can be detected in its early stages, because the bird frequently opens its mouth and apparently gasps for breath, for the thickening of the membrane of the tongue impedes respiration.

INFLUENZA.

Influenza should be treated with warmth; the eyes and nostrils, if a discharge is observed, bath frequently with warm water. Generous feeding should be resorted to, but not of too stimulating a nature, and the bird should be kept by itself. Ordinary cold and catarrh come under the same head.

INFLAMMATION OF THE MUCOUS MEMBRANE.

Inflammation of the mucous membrane is usually evinced by dysentery, the bird mopes, is purged, and the evacuations are tinged with blood. If it is treated promptly directly the first symptoms show themselves, it is possible to effect a cure by giving small doses of castor-oil, and afterwards doses of the *Hydrargyrum cum creta*, rhubarb, and laudanum in the following proportions:—

Hydr. cum creta	3 grains.
Rheubard	2 or 3 grains.
Laudanum	2, 3 or 4 grains.

This dose should be carefully mixed in a teaspoonful of gruel, and should be given every alternate day for ten days, or even longer. The bird must be very sparingly fed, and not on too relaxing food, and for some time after should be subjected to constant supervision. This disease, if it is cured even, always leaves a certain amount of delicacy behind it.

PARALYSIS.

Paralysis arises from over-feeding and too little exercise, and should be treated in the same way as megrims. Soft food should be given alternating with a little grain, but not to much of either, plenty of exercise, and castor-oil in moderate doses. There is little else to be done, as the primary cause is usually derangement or disease of the spinal cord.

PHTHISIS.

Phthisis is the effect of cold and damp; change of climate has been advised by some as the only remedy, but however far gone people may be in the mania of poultry-keeping it would be hardly worth while to pack off a bird to a warm genial climate as would be done with a consumptive patient! Cod-liver oil might be given in capsules in the early stage of the malady, but with me I know it would be a case of *le jeu ne vant pas la chandelle*, and the bird would be speedily be put out of the way as mercifully as possible. For it could be of no possible use if unhealthy, especially from such a cause—no one in their senses would breed from it and so perpetuate the disease, and therefore it would simply be an act of folly, however valuable or well-bred a creature it might be, to cosset it back into a convalescent state and spend time and money on it.

RHEUMATISM.

Rheumatism is caused by cold and damp, birds in very rainy weather in winter frequently suffering from it. It can only be cured by moving the birds affected to dry, warm quarters, giving them nourishing, stimulating food.

CRAMP.

Cramp very often attacks fowls; it is, no doubt, in the first instance attributable to damp, and to stone and brick flooring in fowl-houses. If your house is floored in this manner and your birds begin to suffer, at once have it dug up, and a composition of chalk, gravel, and cinders substituted and rammed down until hard and smooth. If, however, your floor is all you could wish, then probably the place is damp, and there is nothing for it but to seek a fresh situation. The fowls affected will require a good nourishing diet: bread soaked in beer, a little meat now and then, and oatmeal, with coarse black pepper or peppercorns pounded, mixed with it.

Some birds—the large heavy kinds—often appear to be suffering from cramp when in reality their limbs are only too weak to support their heavy bodies; this is frequently the case with young Brahmas and Cochins, and disappears as they become older and stronger. When they are observed to rest on their knee-joints they should be fed with strengthening food, and have iron in their drinking-water.

Cramp in the limbs is quite distinct from leg-weakness, and usually arises from the causes before mentioned, though from the birds being more subject to it in cold, wet weather, they should be well fed and be placed in a warm sheltered place; it is often set down to rheumatism. Sometimes cramp is caused, though by some derangement or obstruction of the rump-gland.

BRUISES AND FRACTURES.

Bruises and Fractures require careful attention; a broken leg may be carefully set by any one clever with poultry, and the bird will do well.

BROKEN-LEG.

The leg can be bound round with stiff leather or tightly fixed in a cork splice bound round with wool. If placed in a shallow basket, with linen cover sewed all round and a hole made through which the neck can move freely, this will leave the leg in perfect quiet, and give it time to set. Food and water can be placed within reach, and three days will suffice for repairing the injury. Let it take exercise little by little. Ducklings spiked by a sharp garden fork in the web of foot or bill will soon recover, as the place heals of itself; if through the neck, bind it carefully round with diachylon plaster.

DROWNING.

No water tubs or large pans of water, buckets etc., should ever be allowed to stand about uncovered in the poultry yard, as young birds especially are very inquisitive, and often fly or fall in. If discovered in time, and any signs of life is left, immerse the bird at once in warm water, dry it, and then wrap carefully in cotten, wool, or flannel, and place near a good fire. Give from half to a whole teaspoonful of port wine, and be careful not to expose again too cold to suddenly.

OBSTRUCTION OF RUMP-GLAND.

Obstruction of rump-gland occasions very often an entire derangement of the whole system, and is nearly always symptomatic of a feverish condition; it brings on inflammation and a tumour, which should be opened by a lancet, and the matter very gently pressed out. Afterwards the place should be fomented with warm water and the bird put on a diet of oatmeal, green food and have a teaspoonful of castor-oil given it at roosting-time every third night until its system appears to have recovered itself. If the tumour is not opened (the incision should be a longitudinal one) the obstruction to the glandular secretion will affect very probably the spinal cord, cause cramp and paralysis in the legs, and end fatally.

CORNS.

Corns are very frequently caused by the birds having their perches fixed too high from the ground; in flying down they bruise their feet, and corns begin to form. In bad cases they should be pared down with a penknife and lunar caustic applied. The best thing to do, though, is to remove the primary causes, the high perch or stone floor, for that will also develop the disease.

ULCERS.

Ulcers do not very often give trouble; they should be opened with a lancet and fomented with warm water frequently. I never had a case, so cannot speak from experience of their treatment.

ELEPHANTIASIS.

Elephantiasis which attacks some of the Asiatic breeds, is the work of an insect, and can easily be detected by the curious growth of matter on the feet and legs; they should be dressed with sulphur ointment, and the birds affected seperated from the others.

BUMBLEFOOT.

Bumblefoot is a malady to which *Houdans* and *Dorkings* seem more subject than other fowls. It is caused very often as corns are, by high perching. The swelling in the foot may be opened with a sharp knife and the matter removed; after this is done the part should be burnt or cauterised. If, however, this is not done, it should be dressed with either common salt, or diluted carbolic acid. Sometimes the fifth toe swells up; this may be prevented by strapping it up to the leg when the bird is about two months old; this will cause it to settle in its proper place and not interfere with the other toes. When you observe your poultry suffering from corns or "bumblefoot" you should lower your perches at once, unless they are already very low, in which case there must be some other reason for these maladies—perhaps a stone or brick floor to the house.

WHITE COMB.

White comb is common in *Cochins*, and is said to be catching. *Hamburgs* also suffer from it. It consists of small white spots scattered over the comb in patches; it extends in some cases to the face, and even into the neck, causing the feathers to fall off. It is not, really speaking, a local ailment, though generally treated with oil, carbolic acid, and such things as if it were; the fact

being that the bird is out of health, and internal heat, caused by too much heating and stimulating food, shows itself externally in these white spots and patches; bad water is also said to cause it, or uncleanliness. The bird is cured certainly after the oil, &c., have been frequently applied; but it is generally Nature's cure, and if no oil had been used would still have been effected. Castor-oil, in teaspoonful doses, is really about the only and best thing to be done.

PAINT AND TAR ON FEATHERS.

In case any of the fowls should get paint on their feathers, rub carefully the way of the feathers with turpentine or benzine; for tar, use butter or oil; but prevention is better than cure.

TORN COMB.

Of frequent occurrence, from the fighting of cocks through wire partitions. Should the wattles be much torn, cut off hanging and jagged parts, and cleanse with cold water and apply cobwebs. When scabs form, zinc ointment will help to soften them.

VERMIN.

Fowls suffer dreadfully, if not looked after, from fleas, vermin, ticks, and other insects. Their houses should be whitewashed at least twice a year, old straw never suffered to remain in the nests, and, above all, they should be provided with *dust-baths* of sieved ashes; they should never be kept without a heap of fine dust or sand in which to roll themselves. If allowed this necessary contrivance they will rid themselves of their troublesome enemies. No one who has watched a hen in a dust-bath will fail to see how thoroughly she enjoys it, sending clouds of dust all over her body and between the purposely ruffled-out feathers; she works her whole body about until every portion of it has come under the action of the bath, and she has completely dusted every part. These baths should be renewed frequently, the best way of doing so being to water the heaps of dust, then scrape the mud so formed together and remove it, filling up the hollows with fresh ashes and sand.

Chickens are very often infested with minute parasites; they should be dusted well with flowers of sulphur, which will generally relieve them of these torments, only it will have to be done frequently to be of any use. Sulphur should also be mixed with with the dust in the baths—flowers of sulphur, of course.

In spring and summer it is especially necessary to keep your fowl-houses perfectly clean, for at those times, of the year insects of all kinds breed and multiply with wonderful rapidity.

The houses should be well whitewashed as spring comes on, and also sprinkled frequently with carbolic acid; size should be mixed with whitewash, as it then fills up the crevices and minute interstices better, and does not easily rub off. Coops should also be washed over in the same way. The dust-baths in summer require to be changed more frequently. In cold weather the various fowl-parasites do not flourish so much or increase so rapidly, and therefore the places will not want so much attention. Fowls will never do well unless they are kept beautifully clean; they will keep themselves so if they are only given the proper requisites, and surely if they are profitable and useful to us we need not mind a little trouble taken on their behalf.

BAD HABITS OF FOWLS.

Fowls have bad habits which need correcting; not so numerous, perhaps, are their faults as those of unfeathered bipeds, but, such as they are, by no means easy to cure.

They are quarrelsome and greedy, but then " 'tis their nature to" be both. *Egg-eating* and *Feather-eating* are two much more serious faults, and unfortunately very difficult to break birds of when they once acquire them.

A hen I once had, a Hamburgh, began to eat her eggs, and though I tried everything I could think of in the way of dosing her with eggs filled with cayenne pepper and various other messes it made no difference; finally I had her fatted up and ate her, which certainly put a stop to her egg-eating propensities.

In those days nests with false bottoms had not been thought of, or, if they had, I had never heard of them. They are the best things to have in such a case. These nests are made with a hole in the middle of the false bottom, which is sloped all round down to the hole.

The nest must not, of course, be lined with straw, which would prevent the egg rolling into the hole, but padded instead, the padding being arranged to slope into the false bottom. Straw or hay must be put beneath the nest for the egg to drop on to through the hole. One can imagine the disgust of the hen who is conscious that she has laid her egg, but yet fails to find it to make a meal on.

Unless, however, the hen was of a very valuable breed, it would be hardly worth the trouble of having such a description of nest made solely for her use. I should be more inclined to make use of her as an addition to my larder.

Feather-eating is also a very difficult thing to cure. Salt is very strongly advised to be given to all feather-eaters. Fowl-breeders use the following mixture, which I consider efficacious: "Four parts of bran, one part meal, one tablespoonful of salt to every eight quarts, stiffly mixed and cooled; birds fed with it twice a day."

Other people advise powdered borax to be given to feather-eating fowls; they will feed greedily on it at first, and then by degrees cease to care for it, or for their feathers. The habit arises, doubtless, from some inward craving. I have known birds in hot weather take to it when the water supply was not sufficiently looked after.

GENERAL TREATMENT OF FOWLS.

If you keep many or few fowls you must be prepared to keep them with some degree of comfort to themselves—that is, if you wish the result to be satisfactory—and to insure this three things are necessary: *good food, clenliness and warmth.* You must have a house for them, and a run attached to it, if they are not allowed to roam about the farm or offices. Fowls kept in comparative freedom naturally do far better than those which have to undergo confinement; they are unquestionably more healthy, for pure air; pure water, and freedom to roam at will over the homestead, picking up insects, worms, green food, and gravel as they like is conducive to a thoroughly healthy state. Unless, however, you have a little land attached to your house, or you are the happy possessor of a small farm, your birds cannot be allowed to follow their own sweet will, and must be kept in confinement at least some portion of the day.

My first essay at poultry-farming was on my father's farm, where, of course, the fowls—at least, all the commoner sorts—had their full liberty. The different special varieties I kept perfectly distinct on the "Upper Farm," as it was called by courtesy; while the commoner useful birds had the run of the farmyard, rickyard, and nearest fields, there being there no garden to spoil. I leave my readers to judge which lot of fowls did best, the pampered, well-bred dirds, who were looked after and allowed their runs in the "mowhay" at certain times of the day, or the mongrel lot who were allowed their full liberty, not being so frequently or so well fed as the others, but having the advantages of freedom to counterbalance these disadvantages.

When kept in confinement three meals a day is the ontside allowance for them, but see that they all get fed, and that the strong birds do not gobble it all up and the more weakly ones fare badly.

Fowls do fairly well in a limited space provided that you can give them a grass run of some sort and you do not keep too many. If only one cock bird is kept, then the hens should certainly not exceed eleven at the very most. Eight is the better number. No cock bird should be kept more than three years. I prefer changing them at two, and two good laying seasons is quite as much as you can expect for a hen bird. After the third year they do not lay so well or so regularly, and therefore, if you keep only a limited number see that they are young and prolific. Directly they begin to drop off in their laying during the proper season draft them off, Either sell or slay and eat! Fowls live to a considerable age. It is, however, a very great mistake to keep on old birds or to breed from old stock, or, in fact, to keep any fowls that are sickly or have passed their prime. Chickens obtained from them are always weakly and degenerate, the eggs slack off, and such birds are not really worth their keep, and if economy is any object should be got rid of without delay. Keeping only the limited number I have advised, you should see that they are *all* in thoroughly good condition. You should manage also to have each year pullets coming on to lay as the older hens begin to cease and want to sit.

Food should be *regularly given* and frequently changed, sameness of diet being by no means desirable. It is equally bad remember, to *overfeed or underfeed*. Study your fowls carefully, and try to hit the happy medium.

My next attempts were conducted on an entirely different plan. In the first case I had the run of the farm produce, had no trouble in procuring "after-wheat," barley, or oats, and plenty of roots and green food. In the second I had to buy *all*, but the scraps from the house. I kept an account of all incomings and outgoing, in a book, entering tne fowls' food, etc., I bought on one side, and the chickens and eggs sold on the other. This plan I advise all poultry-keepers, whether on a large or small scale, to adopt, as it can then be ascertained exactly how much the fowls cost each week, month, or year.

It is impossible to lay down hard and fast rules as to the amount of food you allow. Some days your birds will eat more than others, at other times less, but whatever you do never stint them, for that is a very poor and mistaken economy, as you will in the long rnn learn to your cost.

Poultry-keepers must not mind trouble : look after the fowls yourself. Believe me, personal supervision goes a very long way towards success. Study your birds, their individual ways, their appetites. See that they have all they want or that is desirable for their comfort. Read books and articles on the subject, and treasure up all the *practical* knowledge and hints that you can obtain from those who are experienced in the business of poultry-tending.

DAILY ROUTINE.

METHOD must be observed in every undertaking, more particularly is it necessary in poultry-keeping. You will get through much more work in the poultry-yard by arranging your time, and will have, therefore, more leisure for your various other duties, for I do not for an instant imagine the care of your fowls to be your sole end and aim in life.

The first early meal your cook or kitchen-maid will most probably have to give your fowls, as it is not very likely that you will be up and about at so early an hour; she, too, will let them out of the fowl-house for their early run—that is, if they are allowed and can have the benefit of such a luxury.

Then will come *your* work.

If possible get your fowl-house swept out before breakfast. If you leave it, till after, and you are the master or mistress of a house, you will have that to attend to first, and you will be so late out in your poultry-yard that it will be time for the birds' second meal.

It is astonishing how short a time it takes to sweep out and sand your fowl-house if you do it *every* morning and have everything ready to hand : a hard broom, a spade, a small wooden box made in the form of the boxes grooms use in the stable, in which to place the sweepings and carry them away to the pit. For a pit should always be provided, as, mixed with other ingredients, the manure from poultry is most useful in the garden, but not alone; some people have an erroneous idea that without the mixture of any other substance it is very valuable; but this is a mistake. Then, when your house and yard are clean and sweet, broomed and sanded, see to the water-pans, thoroughly cleansing them, look to the nests if they require fresh straw, and attend to the sitting hens if you have any. A good plan is to let them out for their feed, drink, and dust-bath while you are cleaning and sweeping, as you will then see that they go back properly to their nests again.

Fattening chickens and young chickens will have been attended to before you began your ordinary work.

About eleven or twelve o'clock the fowls should have their second meal—of soft food this time, as grain will probably have been given to them early; then, unless you have young chickens, you need trouble no more about your birds until about five o'clock in the summer and four o'clock in the winter, when you may, if in the former period and very hot, give fresh water in the pans, collect the eggs laid that day and place them in their proper box, and finally give the third and last meal before they go to roost.

Go round yourself the last thing to lock up the house and yard, seeing that all the birds, young and old, are safely housed for the night.

So much for ordinary everyday work; an hour in the morning, or perhaps an hour and a half if you have many small broods of chickens, another hour or so in the afternoon, with a glance now and then into the poultry-yard during the day, will see you well through your daily work, unless there may be, possibly, hatching going on, or sickness, or anything more than usually important.

As to the work being hard, that it certainly is not, for I did it it at one time every day; it is very healthy, and, though possibly to a lady might be a little disagreeable until she got used to it, to any ordinary person it is simply a pleasant outdoor amusement calculated to interest and banish *ennui* effectually.

If you look after your fowls like this yourself, never fear about their succeeding—they *must* do so—and the more care you take, the more thrifty and economical you are in your methods of management, the more you and your poultry will prosper, *only do not begin on too large a scale at first*, and do not throw it all up in disgust because you do not at once succeed in gaining large profits.

Every undertaking must have a certain commencement, and success only follows on perseverance in poultry-keeping as in most other pursuits.

HOW TO COLLECT THE EGGS.

Have a series of shallow boxes (wooden) made large enough to hold as many eggs as you are in the habit of having from your hens every day; these boxes should be filled with bran and the eggs stood upright in them, the small end downwards. Place the boxes on a shelf where you can easily get at them; say you have seven boxes; this seems a large number, but I will show you why; most people date their eggs, but if you have from fifteen to twenty a day you will find this a lengthy process, besides going against the sale of them. If you want to keep the eggs of any particular hen, then date them of course, and add the hen's name; but if you are merely collecting your eggs for table use, or for sitting, you will find this a more simple method.

Your seven boxes should be kept in a line, or if not, on shelves one below the other. You start, we will say, with the first box, in which you deposit ten, twelve, or fifteen eggs, or whatever the number may be of that day's laying. Each box should have a nail, on which should hang a card with the number plainly printed on it, as many numbers as there are boxes, 1 up to 7. On the box in which you put your first day's eggs hang label No. 1. This will be, we will say, on Sunday. On Monday place that days' eggs in the next box; shift the label of No. 1 to it, and put

No. 2 label on it instead ; and so go on until you have filled all your boxes, and they are all properly labelled according to age —one day to seven. No egg is really fit to be called a fresh egg which is more than seven days old. If you keep them longer you must have a store-box and remove them to it, only they should be treated then as preserved eggs for cooking.

HOW TO PRESERVE EGGS.

There are various plans advised by poultry-keepers, but the most simple is really, I believe, the most effectual, at least for two or three months' keeping. The great secret is to close the pores of the shell, which can easily be done by rubbing them well over with butter, oil or lard. The best way is to run down some hog's lard in a basin, and dip the eggs one by one into it. Of course, it must not be hot, only just warm enough to melt the lard and allow of the egg being dipped well in. The greasy substance must then be well rubbed into the pores with the fingers. They can then be stored in bran in your store-tub, or packed in a barrel in hay if you wish to send to send them away. If you keep them in bran they must be looked to from time to time to avoid the bran becoming damp or mouldy, which would spoil the eggs.

The following is a good method for preservation:—"Take a couple of hot lime shells, and in an earthen dish slake them in a good quantity of water, stirring the mixture round to let the sediment fall to the bottom. Put it aside for some hours, and when the superfluous water has risen to the top pour it off till the thick creamy soft lime is reached, for this is the material which, hardened round the packed eggs, preserves them effectually for months to come. Then take the eggs, and in any suitable vessel (chipped or cracked milk-dishes useless for milk answer very well) put them in, a layer all straight on end, and with a spoon fill up with the lime till the eggs are more than covered. Then put in another layer, and proceed as before until the dish if filled up. Put away in a cool but not too dry place— a cellar-floor answers well—and when required for use pick the eggs carefully out from the lime with the point of a knife or a sharp spoon."

Here is another recipe from a German paper:—"On 27th March, 1879, some eggs were placed for an hour in a solution of 50 grammes of salicylic acid and a little spirits of wine diluted with one litre of water, and afterwards were packed away in bran. At the end of June they were in good condition and as fresh as new-laid. Autumn-laid eggs would keep much better than spring eggs."'

The great secret of preserving eggs fresh without recourse to more elaborate processes is to place the egg small end downwards and keep it in that position. This should always be the position of an egg whether kept for sitting or use, rubbed with grease, or preserved in lime and chemical preparations. An inch board about a foot wide and from two to three feet long may be procured, and holes bored in it of an inch and a half in diameter. Strips of board or laths may be nailed round this board as a ledge, and a cupboard in a cool place fitted up with 5 or 6 inch boards or shelves. Then as fast as you get fresh eggs place them in the holes in the boards small ends downwards, and they will keep fresh for some weeks. These shelves might be substituted for the boxes I have before named, and numbered in the same way. A carpenter would make boards of this description for a very small sum.

Eggs can also be preserved by being brushed all over with a solution of gum-arabic, and then packed in dry charcoal-dust; also by being kept in the following mixture, which was invented by a Mr. Jayne, of Sheffield:—In a tub place 1 bushel of quicklime, 2lbs. of salt, ¼lb. cream of tartar, and mix all together with sufficient water to make the composition of such consistence that an egg put into it will swim with the top just above the fluid; then put in the eggs, which, it is said, will keep good for two years.

HOW TO PACK.

Each egg should be wrapped first of all in silver paper, then newspaper, and then placed in a box filled with hay, rough seeds, or chaff; there is then no fear of their breaking. It may sound troublesome, but if silver or thin paper is always saved, and newspapers, and placed aside in a drawer or shelf devoted to that purpose, I see little difficulty about it myself. It is harder to procure the seeds or chaff. If these cannot be had, then soft hay or straw must be used, but remember the secret of good packing is tight packing; leave no crevices, or else the eggs will shake about.

In packing sittings of valuable eggs the above is the best plan; but, as a final precaution, the box so packed should be packed itself into a hamper well surrounded with tightly-pressed hay or straw. I may mention that the lid of the box ought to be *screwed* on for eggs which are to be used for a sitting, as the jar, in nailing, would very likely damage them. I generally use hay myself for packing the wrapped-up eggs, and I never heard of any damage occurring to them. Many people, I know, advise chaff or seeds; but sometimes—in fact nearly always—the shaking of the journey causes the seeds to settle down more closely together, and then spaces may be left, though you might fancy you had packed your box perfectly tight. With the hay this is not at all likely to occur, especially if you put a good layer of it in the bottom of the box, then a layer of wrapped-up eggs, then more hay, and so on until you have filled the box.

MISCELLANEOUS
Poultry Receipts.

ADDLED EGGS.

Those which have been sat upon, and which have been clear from first day, or "unfertile." Such eggs, if taken out of the incubator or nest on the fifth or sixth day, are fit for cooking; if left in, they will injure those that are fertile, and eventually, when broken, may explode and be very offensive; if incubation is continued with a *"clear"* egg it will become "addled." An egg becoming fluid shows that at some time it must have been chilled.

AGE OF HENS AND PULLETS TO LAY.

Pullets lay at four and a half months to six and a half; their time for laying is in the hands of the clever poultry-woman or breeder. If hatched for egg produce only, and with intent to keep up a succession of layers, eggs should be incubated during almost every month: January to April being the best. If your object is to secure eggs at the earliest moment, take care not to move the pullets about from place to place, as it seriously delays and checks egg production; as soon as pullets are over their fifth month they should be mated, and, if possible, in the same quarters in which they have been brought up. Chicks hatched in September will not be so forward to lay at five months in January as those hatched in January are in April. Hens, if they are old and have laid on late into the season, cannot be expected to lay before February.

AGE TO BREED AT.

Pullets, on the contrary, meant for exhibition and prize stock must be hatched in the months of January to March inclusive, and it is wise to give the pullets a frequent change to prevent laying, and so lengthen the period for growth. Pullets from which to breed exhibition stock should be quite seven to eight months old before they are mated, and when this is done an adult cock should be in their run, while the cockerels should be placed with the adult hens. Hens from two to three years of age are more valuable to breed from than pullets: their chickens are stronger as a rule.

AGE TO FATTEN.

Cockerels of from four to five months are in prime order for fattening; it is little use commencing earlier if size is an object, but birds of this age should be ready for table if well kept without special fattening.

AGE, TO JUDGE, IN POULTRY.

If for table, examine the feet and legs; the size and appearance of the spur form an unfailing guide in hens as well as cocks. The skin of the pullet or cockerel is smooth and fresh-looking, the adult bird yearly grows coarser and shrivelled: place the thumb and forefinger on either side of the back near the pope's nose or oil receptacle, and press it; in young birds that part is supple, in old birds it is difficult to bend. In ducks the age is very hard to tell, even great judges find a difficulty; ducks of two or three years' standing have, however, a deep depression down the breast feathers, and their waddle becomes yearly more ungainly.

AGE TO KILL.

If you are breeding for exhibition kill off any birds not fit for show purposes, HOWEVER SMALL, unless you have special runs for table stock into which to draft all false-feathered specimens. It is bad economy to keep them crowding your prize birds preparing for exhibition. Hens should be killed in their second year, the moment they cease laying their summer batch of eggs; do not let them get into moult, or you will have to keep them through it at a disadvantage. A hen hatched, say, in March, 1878, should lay up to the end of July in 1879, and then be fattened and killed; if well hung, she will be excellent eating still. In case of prize stock the rule is different, and the moment for killing must be decided by the value of the specimen in question; no valuable stock bird should be discarded till the completion of her third or fourth year, as even a dozen or two of her eggs may be invaluable for incubation.

AGE TO SHOW.

The exhibitor who wishes to be early in the field must hatch early. Cockerels for August and September shows must be hatched out in January and February. In from six to seven months the plumage of the cockerel is fairly perfected, and pullets, if kept from laying till six or seven months and then paired, and exhibited when laying has just commenced, will be in perfect order. Adult birds must be exhibited when over moult, and when laying re-commences; the cock bird is not in spirits and will mope in his pen if shown before the moult is over, and the adult hen's comb will look dry and shrivelled up till she lays.

AIR BUBBLE, OR AIR CELL,

In an egg, is at the round or blunt end: it contains the air which is to supply the chick during the process of incubation—it is known to contain a greater proportion of oxygen than the air we breathe. As the chick increases in size the air bubble grows larger, and when on the point of hatching it occupies one-fifth of the whole egg. The slightest perforation of this air chamber will prevent succes in hatching out. About the nineteenth day of incubation the air cell is ruptured, and the chick breathes with its lungs; it is at this date that the lively movements seen in eggs placed in water are observable.

ALARM.

The attendant in the poultry-yard should by all means try to make friends with his flock, and never alarm them or drive them needlessly. He should never shout at them in an angry tone, and should acquire a habit of calling them always in a special manner when coming to feed them at meal hours. When laying hens are alarmed soft eggs are the result; and if afraid of their keeper, exhibition birds in being caught and dressed for shows run the greatest risk of spoiling combs and plumage by knocking themselves about in their fright.

ALE.

Ale may be used with advantage occasionally as a treat in severe weather for laying stock. Should a bird's comb turn blue, from confinement at a show in cold weather, a warm house and bread soaked in ale once a day will effect wonders. But ale should not be used constantly, as it is too stimulating.

ALLOWANCE OF FOOD.

(See Food.) No absolute scale can possibly be given, and yet on the proper allowance of food being given to each fowl hangs the question, "Does poultry pay or not?" Most fanciers' birds are over-fed, and in this lies great danger. The various breeds require more or less food according to size and character, and the only safe rule to lay down in case of adult birds is to give at the morning meal and supper time only as much as they will pick up greedily, (never leave food on the ground or in the pans) at mid-day give only a few grains, more as an occupation than as a meal. I find a (lady's) handful of grain to each bird is ample; while of soft food in the morning, about as much as would go into a gill measure is almost too much for each bird. Green food should be given *ad lib.*, and should never be forgotten, but too large a supply of fresh-mown grass given at rare intervals is injurious; birds are apt to eat too much, and scouring is the result.

ANIMAL FOOD.

When fowls are kept without grass run or ample range, meat is all but a necessity; the number of eggs in winter will be greatly increased on meat diet. Bullock's liver, slightly boiled, and chopped in chopper, peppered and mixed with pot-liquor, and thickened with oatmeal, warm, is much relished. Scraps and cuttings from the butcher, are well worth getting for the same purpose. The French give horse-flesh to their poultry, in the management of which they are notably successful. During moult and in severe weather it is bad economy not to buy meat; for breeding stock and early supply of fertile eggs it is essential. It may be given chopped small and mixed with rice, meal, or other soft food; or, for a change, a piece of meat may be nailed or tied to the wall of the pen for the fowls to peck at on rainy days. For chicken supply, cook the meat and chop it fine. This may be given to chickens at three days old and upwards. A supply of maggots is the best animal food for young chickens. If in the country, a dead lamb or sheep, or a calf' may at times be procured, and provided that death was caused by anything but a long and severe illness, there is no reason why these should not be used in the poultry-yard. Attentions of this kind in ₁ poultry-run ward off evils, such as egg and feather eating.

ANIMAL FOOD: KINDS TO PROCURE.

Meat: Gravy-beef, slightly boiled, salted and peppered and finely chopped, should be got.

Earth-worms are excellent, but some fowls will not eat them.

Scraps from table.—Very good; and if near a boarding-house or school, arrangements should be made to take the remnants.

Ants' Nests.—Excellent for young chickens. School or village children may be encouraged to bring them in for a few pence.

Meal-worms.—These for chicks are unsurpassed, and produce wild excitement. A regular supply can be procured by expending a little trouble and thought on the matter, which is, however, not an agreeable one.

Cheap Winter Food.—Get a sheep's or cow's head; boil it down, after chopping it small with an axe or a bone-crushing machine. Boil in the stock, cabbage, potatoes, carrots, parsnips, and thicken to a crumbling mass with dry boiled rice.

APPETITE.

To maintain birds in a really healthy state appetite must be kept up, and it is good management to have the poultry in such a state that they will fly up to meet the poultry-man, and scramble for their food. Loss of appetite comes from unwise feeding on over-spiced food.

APRIL WORK.

March chicks are always more in demand than later ones; but April is the month when the birds get on best, and hatching should be vigorously continued. If warm and dry, nests should have very hot water poured round the outer edge of the straw—not on the eggs; avoid doing this so that the water runs into the nest. The eggs should not lie in the water for a moment.

Drench all the parts of the nest except the eggs.

Chicken feeding becomes complicated; the older broods of January and February will be growing large and very bold. If it has not been done already, separate them; draft off the older broods to open quarters as weather permits, so that younger chicks may have a better chance. Place the food about in a number of glazed saucers or tins, so that all may have an equal share. Let all eat until they will eat no more, then clear away the remains.

Continue constant disinfection of chieken boxes and coops; eschew hay beds—they breed vermin. If a brood looks sickly, suspect vermin—"lice"—and treat accordingly.

Show cockerels and pullets should now be penned up separately and receive especial care. Commence to fatten early hatched January chicks not fit for show or stock. Feeding the young chicks at 10 p.m. may now be abandoned, but soft sweet food and handfuls of grain should be placed about in the evening after the chicks are gone to bed, to be ready for the early morning.

ARTIFICIAL INCUBATION.

Practised from the earliest times in the East chiefly in China, India, and Egypt. In the latter country large *mamals*, or ovens, holding from 40,000 to 80,000 eggs, are still used for the purpose; and the villagers bring their eggs in the expectation of receiving after a lapse of 21 days, 200 chicks for every 300 eggs deposited. In 1851 Cantelo exhibited a hydro-incubating machine in London, and the subject of artificial incubation has occupied much attention in France from that time to this. Rouillier's and Voitellier's machines have there taken the lead, while quite recently in England the matter has been taken up with more or less success by several competitors for public favor.

The chief desiderata in artificial incubation are a regular temperature—from 100° to 106° Fahr. at the upper surface of the eggs—moisture, and after the tenth day adequate ventilation (without chilling) of the eggs. The source of heat is immaterial so long as these essentials are secured; pratically, however, a lamp, or gas, and boiling water are utilised for the purpose.

The choice between a gas or lamp machine and one worked by boiling water will be probably determined by the greater or less facility with which boiling water in quantity, say two or three gallons, night and morning, can be obtained. Excellent results have been secured by both systems, and 75 per cent. and upwards may be hatched out by either method with reasonable care and attention to detail. The chief points to be attended to in artificial incubation are:—

i. Position of incubator: an out-house, conservatory, or room on the ground floor is best, where there is an equable temperature, good ventillation, and no risk of jar.

ii. Let the eggs be as fresh as possible: allow travelled eggs to settle for twenty-four hours before placing in the incubator.

iii. Avoid putting eggs in the drawers by twos and threes, as far as possible; if unavoidable, warm the eggs well at 104° to 105° Fahr.

iv. Aim at a regular temperature of 103° increasing after 9th day to 104° Fahr.; with high-bred eggs a degree higher is an advantage towards the close of hatching. Occasional extremes of 90° Fahr. or even 108° Fahr. have been borne with impunity, but safety lies between 100° and 106°. Over-heating is more dangerous than under-heating.

v. Clean the eggs well; mark them on one side with date, &c., rather nearer the smaller end, and place the eggs date downwards, for the first twenty-four hours; your marks will then be preserved when the chick is hatched.

vi. Keep the trays of earth or sand always moistened beneath the eggs, but avoid sprinkling the eggs themselves with water.

vii. Turn the eggs slightly every twelve hours. Air them during the first ten days for ten minutes at night and a quarter of an hour in the morning; for the last eleven days allow twenty minutes in the morning and fifteen minutes in the evening.

viii. Sixth day examine all eggs, and reject those which do not show embryo and distinct veining; the clear eggs are still fresh for cooking even for custards.

ix. After the ninth day air is required by the chicks, but chills must be avoided, especially when nearing the last stage of incubation.

x. When eggs at different stages of development are in the same drawer, it may be well to reserve one portion for those last put in, and to cover them with a flannel, whilst those in a more advanced stage are aired more freely.

At five or six weeks, according to the weather, chicks do better without artificial heat, and a run on grass or dry earth beds when the sun is shining will be found most beneficial to health.

A late vinery with earth floor is one of the best places for rearing early chickens. Great care, however, will be required if many are bred, to keep the earth fresh dug. Cleanliness also is essential, especially in the mothers, which should be well aired daily, as well as being provided with a fresh earth floor. A little disinfectant, such as carbolic acid or terebene, will often be a useful addition.

ARTIFICIAL REARING.

The chick, after leaving the egg in an incubator, may be either placed with a hen or be brought up under one of the many artificial mothers which have been invented for the purpose. In some machines the mothers are attached to the incubator. A hot water bottle suspended by flannel in an open box forms a cheap and useful arrangement for infant broods. A well-sustained heat of about 80° Fahr. is essential for the first week at least, as a chill is fatal; after that time it may be ten degrees lower. The run should not be large at first, and should be protected, by wire or netting, from the access of older birds. The broods should be carefully sorted under different mothers as to size, and the greatest cleanliness insisted upon. For twenty-four to thirty-six hours after leaving the eggs the chicks requires no food. It may stand in the midst of plenty, but it will not peck or do any-thing but pipe. If healthy, it will begin to pick when hungry; a little tapping on the floor, however, or showering crumbs on the backs of the chicks will excite the more backward ones to help themselves.

The first food should be hard-boiled eggs, chopped fine and milk, to which may be added fine shreds of lean meat or bread-crumbs.

ASIATIC BREEDS.

Comprise the various kinds of Cochins, the light and dark Brahmas, Malays, Langshans, &c. They have, as a rule, feathered legs, and lay rich chocolate or yellow eggs, small as compared with the size of the birds. They bear confinement well, have no taste for roaming, and are of a phlegmatic nature. They lay in winter, but the supply is limited, owing to their constant desire to incubate. Of a contented turn of mind, they are willing to keep within the limits marked out by a wire fence two feet high. As mothers they are devoted, but exceedingly clumsy, and woe betide the newly-hatched chick caught beneath the weighty tread. One great advantage of their Asiatic

WHITE CRESTED BLACK POLISH.

fowl as hatching-machines is that they harbour no lice or vermin; they are gentle and easy to handle. Chicks of these breeds feather very slowly, especially if hatched late in season, or the produce of young pullets' eggs.

ASIATIC DISEASES.

Various liver derangements, over-feeding being the cause; Scaly Leg; Apoplexy.

ASPECT OF RUNS AND PENS.

Build them as "lean-to's" against a wall where there is a chimney running up, constantly in use if possible; this is far better than stove heat, and let the aspect be south. Catch every ray of sun. Keep the roof as high as possible in front, so as to let in all the light; shelter by judicious use of boarding from wind, and, above all from cold draughts.

ASPHALT.—FLOORING.

The original soil is as good as anything for the flooring of a poultry-house. If pulverised and screened lime rubble from old walls and buildings be thrown on the earth, and all is kept clean, and well dug over after cleansing twice a week, it makes a soft, dry and deep bed of the best possible dusting material; but asphalt is also useful: it keeps out rats, and, if covered with six inches of sand and rubble screenings mixed, forms an excellent flooring; it must, however, be entirely cleaned out every two months or so, and fresh dusting material supplied.

ATROPINE.

Used in severe cases of roup. Apply a drop to the eye if matter begins to accumulate.

AUGUST.

Many early birds will be making their adult plumage now, and it will be possible to estimate which are the specimens to gain credit in the show-pen. Proceed accordingly, and remove such birds into the preparing pens. Each cockerel should have one to himself, with ample room, or if dull alone, put with him a known companion younger than he is, but not one for show, just to keep him company. Kill or draft out of the way all wasters, and devote all energy to the prize birds and those which will come for sale as stock birds.

Pullets can be kept in *small* flocks, as they never quarrel, but not more than ten together, unless, the enclosed covered runs are very extensive. Prize birds and stock must not be allowed to scuffle and scrimmage for their food like farm-yard mongrels; full diet is necessary, and fighting must not be tolerated when birds are bred for feather, weight, &c.

Early cases of moult in cocks should be seen to, and the birds isolated where hens do not worry. If sudden moult and general loss of feathers takes place there is no cause for anxiety; it is better then a tedious moult later in the year.

Continue weeding and killing; the more room gained the better.

Take steps to procure ducks' eggs to hatch in October. They are always difficult to procure at that time, and it is well to bespeak them beforehand.

AWARDS.

The young poultry fancier should not be discouraged if he only gets honourable mention; true, it does not help, as do prizes, to pay the entrance fees and carriage, but it has been said, with justice, that a run which produces specimens which win honors wherever shown is often better than one which gains several prizes with, it may be, the same well known bird. Let him make up his mind to be beaten, dress his birds with care, and send the best he can; and then, expecting a beating, victory may come as a pleasant surprise, and defeat will not affect him otherwise than to urge him on to "try again."

AYLESBURY DUCKS.

Hardiness, great size, and early maturity are their merits. They have now formidable rivals in the "Pekin" breed which exceeds them in apparent size, but does not beat them in solid flesh weight.

Their plumage must be spotless white, yellow legs, and beaks of flesh color; weight of adult birds 6 and 7 lbs. They have been exhibited up to 20 lbs. the pair.

For breeding, running water is the best, and failing that, a pond is indispensable. They commence laying in December. The color of the egg varies from green to cream color.

At six weeks ducklings should be 4 lbs. each and fit for table; the ducks should always have gritty substance put into their water, wherewith to purify their bills and help them to digest their food; this is of great importance. Ducklings for exhibition may be brought up with more water to swim in than those kept for fattening; their houses should be well bedded with clean straw. Feed at first on chopped eggs, milk, oatmeal, warm; then with oatmeal, varied with barley-meal, given as often as they are willing to feed. As to green food, the supply should be unlimited. It is not well to let food remain in the pans; however often you feed the ducklings, remove what remains, so that they may have an appetite when the next meal hour arrives. It is necessary to give ducklings the means of collecting worms and slugs; the moment they are old enough to roam for this, their natural food, they will be the better for it. Dry bedding is essential, and warm night shelter; otherwise ducklings will not bear the cold of early spring days without injury.

BARLEY.

Good uninjured barley should be a staple article of diet in the poultry-yard. Buckwheat is preferable, but not always to be had, and barley is a safe and favorite grain. Damaged cheap grains must not be ventured upon. Barley warmed in a tin or pipkin in the oven, dry, and given hot, as an evening meal in cold weather, is very wholesome; and for a change, if with water added and a little pepper it is baked till it swells up to twice the original bulk, it forms a much relished dish.

BARN-DOOR FOWLS.

To the present day many amateur poultry lovers continue in the old and popular error, that barn-door fowls "pay" better than pure breeds, and are less trouble to rear. "Fancy" fowls are, without a thought, voted useless, whereas many poultry-yards fail of success owing to the mistake of starting by buying up a lot of mongrel birds, age unknown. Every distinct breed has some leading characteristic, for which in the past it has been selected and bred: it is prized, and, with much care, bred for either size or hardihood, for laying qualities, as good mothers, or as super-excellent table fowls; while the barn-door is not remarkable for any speciality, neither producing so many eggs as the Andalusian or Leghorn, nor being so well flavored as the Game, nor so white and full of meat as the Dorking, the Houdan, or any of the table breeds, nor so content with small pens as the Asiatics: in fact, even from a commercial point of view they are not to be recommended, and are a mistake. For beauty certainly no one would select them.

BEDDING FOR CHICKS.

Straw put through the chaff-cutting machine is undoubtedly the best, but on no account give it uncut, or it will be the source of broken legs. Care must be taken to remove the soiled straw and to replace it with fresh every morning. A small quantity changed daily is far better than a deep bed left uncleansed for a day or two. Cleanliness is the first consideration; sprinkle the bed with powdered sulphur three times a **week, to keep away** vermin. No bedding should be given to birds put up for fattening. The coop in this case should have a barred floor, so that the dirt shall not rest upon it. Otherwise, in such a small space the birds will quickly suffer from bad air and dirt.

BENZINE.

Good for cleaning accidental paint off the feathers, which it will not injure.

BLACK HAMBURGHS.

Of the Hamburgh tribe the best, being larger, and laying an egg of more useful size. They are non-sitters, being crossed with Spanish; white face crops up, and is a difficulty in breeding. Chicks, when hatched, are from the throat downwards to the under part of the body white, the rest black. It is only after the second moult that their glorious plumage of green black glossy hue is perfected. These birds are delicate; damp, dirt, cold, and improper food bring on liver disease. They do not thrive in confined pens, and it is unwise to attempt breeding them unless you can give a grass run.

Points.—Beak black or horn; comb, face, and wattles rich red; deaf-ears white; eyes red: legs blue lead color; plumage deep black, brilliant gloss of metallic green or bluish purple: the greener the better; comb double and evenly serrated.

BLACK ROT.

A disease to which Spanish fowls are subject, but not often seen now. The comb becomes black and the legs swell, with general pining and loss of spirits, flesh, and gloss. Calomel and castor-oil should first be given, and then generous diet, with tonic and warmth, may pull the bird through.

BONE-DUST.

Very beneficial for the feeding of growing birds up to five or six months of age, a preventive of weak legs and diarrhœa; an aid also in postponing the development of young birds, while it provides materials needful for continuous growth, and gives strength and size to the frame. It should be about the fineness of coarse oatmeal, and should be sifted into and with the meals used, in the proportion of 1 quartern to the cwt. Fresh bones chopped and pounded, or burnt bones, are not so useful for the above purposes as they are for laying stock or for birds of an age for exhibition.

BRAN.

A few handfuls should be mixed with soft food, as it is decidedly wholesome, though not a favorite poultry diet.

BREAST-BONE CROOKED OR BREAKING.

This is caused by perches being placed too high from the ground when the roosting-place is of small size. Objections are made to letting chickens roost early, but little harm will come of it if the perches are at a proper height. It is the perpendicular sudden flight to the ground in a confined space which injures and breaks, or bends, the tender breastbone. In limited space, and with chickens brought up in confinement, shelves, sanded, and then littered with chopped straw, are safeguards against this evil.

BREEDING COCKS.

Must be vigorous during December, January, and February. Feed them extra well, and alone. They should be examined when perching at night, and if poor must be at once fed up; they should not exhibited, and the breeding-pens should be made up by Christmas time, as the cocks become attached to their special hens, and changes affect the fertility of eggs.

BREEDING IN AND IN.

Less dangerous than a constant introduction of fresh blood into an already-formed prize strain; but judgment must be used, and in a tolerably large establishment, where the hens are kept away from the cocks during moult, and re-mated in December, arrangements can be made not to mate too nearly-related birds, without recourse yearly to purchases from a fresh yard. It is well to keep a spare cockerel or two in stock, and let them run on some farm, and re-introduce them the following year into your breding-pen, when the relationship cannot produce evil results. If only one pen or yard of poultry is kept, the results of mating the produce from it may be most unsatisfactory, and a fresh cock becomes a necessity.

BREEDING-PEN FOR DUCKS.

Four ducks are ample to one drake. Ducks may be bred in a small pen, where there is merely a tank of water 4 feet by 2, and 2 feet deep. Many of the eggs, however, will be unfertile. For success on a large scale a good-sized pool of water is essential. The birds should have the run of a field. Care should be taken not to overfeed breeding ducks; and if early eggs be required meat diet must be given, and warm quarters at night.

BREEDS FOR CROSSING.

Never cross with a bird which is already the production of a crossed breed. First crosses are the most satisfactory, and take care that the cock chosen for the cross is full-sized, symmetrical, short-legged, with fine bones, plump breast, and strong constitution. The cock of the walk should be pure-bred. White Dorking with light Brahma, or silver-grey with dark Brahma, produce good large birds, and well fitted for fattening for the table, while the Dorking improves the Cochin's laying power. For laying purposes, the Brahma crossed with an Andalusian cock cannot be surpassed, and a propensity to sit is added to the Andalusian's laying qualities.

The Houdan crossed with Brahma or Cochin matures early, and becomes a prolific layer. For table use the dark Brahma and Creve-cœur are good. The Creve-cœur is excellent as a layer, and its size for table and early maturity can hardly be surpassed; but constitution and hardihood may be added by the Brahma or Cochin cross. The flesh of Malay and Game fowls is said to be of super-excellent delicacy. To increase the amount of flesh cross with grey Dorkings, when an excellent plump table fowl will be produced, the couple weighing when fed up from 16 lbs. to 18 lbs. The advantage of a cross is that the produce is usually hardy and larger than the parent birds.

BROKEN BONES.

Machines may be had for breaking bones from the kitchen into small pieces. Such fare is excellent for the growing stock, and will encourage laying in the adult birds; but it does not take the place of bone dust, mixed with meal, which retards development and forms bone and framework, while fresh bones stimulate laying powers, etc.

BROODY HENS.

When broody, *i. e.*, wishing to sit, hens go about clucking for several days, sit longer and longer on the nest after laying, cease laying finally, and do not leave the nest. If a sitting hen is not required, remove her at once to a fresh run and new companions. Shut her out for a few days where no nests may tempt her. If, on the other hand, she is required to incubate, encourage her by false eggs in the nest, and partially protect the entrance to the nest from other prying hens. All Asiatics are much given to sitting, and Dorkings are good mothers. No hen even crossed with Spanish, Leghorn, Hamburgh, or Polish blood will incubate satisfactorily. The broody hens should be fed once daily on sound grain, some grass or lettuce, and a treat of scraps; soft food now and again keeps her in better condition than an exclusively grain diet. On no account deprive the broody hen of her **dust-bath, and if your brood is valuable, take the trouble to** dredge her under wings, legs, etc., with powdered sulphur.

BUCKWHEAT.

Buckwheat should be a staple article of corn diet. The color is a drawback till the birds learn to recognise it as food, when it is greedily devoured. It is not fattening like maize, but good for stimulating egg production. This grain, as prepared in Russia for the use of lower classes, cleared of the husk, and split up into coarse grit, is invaluable for feeding chickens; it is devoured greedily at three days old, and eaten in preference to the best cutlings or grits. Baked in the oven with water, it makes a most valuable soft food, and is much relished by old and young.

BUENOS AYRES DUCKS.

Called also Black East Indian, Labrador, or Black Brazilian. For prize-winning they must be very small, with neat round heads, short bills, short bodies; plumage pure black, with brilliant lustrous sheen of green all over. This metallic green lustre cannot be too great, or the size too small. The bill olive green, the legs and feet black. Easy to breed, for if you do but start with a good pair, the produce will be good; but the smaller you breed them the more delicate they become, and less prolific. They are most excellent table birds, of delicious flavor, and are often bred for this purpose in numbers, though they are also great favorites in the exhibition pen.

BUTTERMILK.

Buttermilk is excellent for poultry; but do not let it stand in the sun, as it is very unwholesome if sour.

BUYING POULTRY AND EGGS.

Buy none to mate with prize strains without careful investigation of the antecedents of the yard and strain. Have the birds on approval; even if the carriage both ways costs $3.00 it is well spent if it saves an introduction which may bring nothing but trouble, and spoil the strain. Strangers to your poultry run and your breeding plans may, without any dishonesty or unfair dealing, send what would ruin your plans. Therefore see your birds before you buy. Be careful how you feed newly-bought birds; give soft food and sparingly of dry grain at first, gradually increasing the quantity; eschew maize and new wheat, which may bring on various troubles. In buying eggs for incubation *early* in the season, make special terms as to the date by which eggs ordered are to be delivered.

Do not purchase eggs for incubation in the two first months of the year without some arrangement about "clear" eggs being replaced by others at one-half or one-third price. Many eggs are then unfertile, and most breeders will replace at half-price if a decent number of chicks do not hatch out. See that your eggs, if valuable, come from known respectable breeders, remembering that one good shake given to the egg when bringing it from the nest may ruin every hope of successful incuabtion.

CABBAGE.

A useful vegetable in winter for poultry. It comes fourth on the list. Even the outer leaves and stems will be greedily picked up if they are chopped up in a chopper. Stalks of cabbage should not be left about, as they give a very offensive smell when rotting in the wet and sun. Whole cabbages nailed or tied to the walls of the pens give amusement to poultry in confinement; there is no better preventive for feather-eating, which is often caused by the idleness of an imprisoned life.

CANKER.

An ulcerated state of the eyes and head; the disease sometimes attacks the mouth and throat. It comes from neglected cold, and damp, improper housing. No well-ordered poultry yard should suffer from such serious ailments. They may be relieved, however, by washing perseveringly with chlorinated soda diluted with four parts of water, and dabbing the throat and tongue with the pure solution. Put sulphur in the food, do not expect immediate cure.

CAPONS AND CAPONISING.

The weight and delicacy of birds are greatly increased by this operation, and it is only in England that birds are fattened for table witnout having recourse to it. The proper age for the operation is four months; it is not considered dangerous—about one in forty succumbs. Capons will remain tender for table up to eighteen months of age, and may be utilised during that time by bringing up broods of chickens, to which they become as devoted as mothers.

CARBOLIC ACID.

One of the best. Prepare thin lime-wash, and into two gallons put two ounces of acid crystals, with which let all be whitewashed. On a real hot summer's day, if the poultry pens are crowded and any fear of vermin should arise, it is well to turn the birds into the open grass run, and with a fine rose water-pot to sprinkle the floors, nests, and perches, after thoroughly sweeping all out, with the same mixture made of double strength.

CARE OF EXHIBITION BIRDS AT HOME.

Protect their plumage from every risk of injury through rough or broken wires, from high perches, from shelves too near the wall or roof of the roosting-house; see that your trap-doors are smooth at the bottom, and large enough to admit adult birds without touching their sickle feathers; never let the pen remain uncleaned even for half a day. Give plenty of clean straw, and see that the dusting material is quite clean and dry. Do not let birds be exposed to a fierce sun, and never to rain. Prevent any fights or sparring between neighbours in pens. Give all the run and freedom you can in fine weather. Feed well with the best and most nourishing food. Give meat and green food, but beware of excess, and never leave food on the ground. Give iron in the water, it produces brilliant combs, and is a tonic; avoid species and condiments. Hemp now and then, linseed boiled to a jelly and thickened with Spratt or oatmeal three times a week for about three weeks before showing, will add to gloss on plumage; the last two meals before showing should be rice boiled stiff in milk, with meat chopped up in it. Prepare your bird for the confinement of a small pen and its trials by gradually penning him up, and getting him accustomed to small quarters, and never send the birds off crammed with hard grain for a long journey. Get the pair to be exhibited acquainted with each other during the last twenty-four hours, lest they should fight in the hamper and disfigure their combs; but do not pen the exhibition pullet with the cockerel till the last day. Rather give him any other bird as a companion, so as to accustom him to ladies' society, and save the wear and tear of plumage to the show pullet in a small pen.

CAROLINA DUCK.

These birds, with their brilliant plumage, bear confinement well, and will breed satisfactorily under favorable circumstances and when used to their home. They should be about the size of widgeons, and when exhibited brilliance of plumage wins the day. The drake should be red in bill, margined with black; orange-red eyes; head bronze-green shading to violet, and a remarkable line of intense white runs from behind and over the eye, mixing with the long green-and-violet colored feathers of the crest. The neck and collar white; breast claret, speckled with white, size of specks increasing downwards. The sides of the body under the wings are marked with black lines on yellow, and the flanks with stripes of black shaded white; tail-coverts black tinged with yellow. The back is a bronze with greenish hue, while the wings are adorned with spots or marks of blue and green. Legs and feet a reddish-yellow. The duck is much more sombre than its mate; crest smaller; the eye-bar. chin, and throat all white; head, neck, and breast shading from dark drab to brown, spotted white; back and wings glossy bronze, with gold and green reflections, and less brilliant wing. spots than the drake. All these brilliant colors and strong con. trast become more vivid with advanced age.

CATARRH.

The moment cold in the head exhibits itself by the usual signs of sneezing and running at the nose, the bird should be removed to a dry, draughtless place, nostrils washed with vinegar and water, giving every third day two teaspoonsful of caster oil; diet—oatmeal and bread in hot beer, and plenty of grass. If not soon cured.

CHANGE OF PLACE.

Poultry are truly domestic and love their homes. If eggs are an object, it is most important birds should not be moved from pen to pen, as it will delay egg production, and diminsh it greatly. Pullets for early laying should, if possible, be brought up in or within sight of their future laying run, or pen, if, on the other hand, it is wished to delay the laying of a pullet, and to encourage growth for prize purposes, her home *must* be changed often. A sitting or broody hen may be best cured by removing her to a new scene with fresh companions—a more reasonable and humane way of checking her maternal instincts than that of half drowning her, or shutting her up in darkness or dirt.

GENERAL TREATMENT OF CHICKENS.

During the first twenty-four hours give no food, and remove, till all are hatched, from the hen or incubator to a box, having ventilating holes bored in the side, and a hot water-bottle slung, by means of coarse flannel, so that the chicks may feel the warmth and the least pressure on their backs. When all are hatched, cleanse the nest completely, and well dredge the hen's body with sulphur powder; give her the chicks, and place chopped egg and breadcrumbs within reach. The less they are disturbed during the first two or three days the better. Warmth is essential, and a constantly brooding hen is a better mother than one which fusses the infant chicks about and keeps calling them to feed Pen the hen in a coop and let the chicks have free egress. The best place to stand the coops is under sheltered runs, guarded from cold winds, the ground dry, and deep in sand and mortar siftings. Further warmth is unnecessary if the mothers are good; and if the roof is of glass, so as to secure every ray of sun, so much the better. Cleanliness of coops, beds, flooring, water-vessels, and food-tins must be absolute. The oftener the chicks are fed the better, but food must never be left; water must be made safe, or death from drowning and chills may be expected. The moment weather permits, free range on grass for several hours daily is desirable, but shelter should always be at hand.

Diet.—The longer the supply of hard-boiled eggs chopped fin. is kept up, the better. As the birds get on, every kitchen-scrap is invaluable, and the following mixtures may be given for meals in turn as convenient, variety being essential for success, 1st meal: as early as possible—6 a. m.—egg chopped, mixed bread-crumbs. 2nd meal: Spratt and a few grains of Scotch groats. These can be brought at the corn-dealer's, their place being taken by good wheat as birds get on, or by crushed barley. 3rd meal: kitchen-scraps chopped fine in a chopper, given warm, and mixed to a crumbling mass: 4th meal: rice boiled in milk, and dried up crumbly with Scotch oatmeal. 5th meal: barley-meal mixed crumbly with the liquor in which meat has been boiled. 6th meal; meat chopped fine and pollard reduced to crumbs (not necessary daily). These preparations given in turn and with judgment will, with occasional handfuls of small, dry grain, and barley and buckwheat baked with water in the oven, give the chickens all that is necessary for building up the strong framework which is so essential to a finely-developed bird. The use of bone-dust must not be omitted, and a constant supply of green food, together with mortar, oyster-shell, gravel and all manner of grit and dust should be ensured. Pure water, not left to stagnate or freeze or to get hot in the sun, and, if possible, milk occasionally, will render the diet perfect; chicks so kept, the quantity given being increased with their size and appetite, will be found at four months, or, at any rate, at five, to be fit for table without the unhealthy and unpleasant process of cramming; if destined for the show-pen, they will be ready to "go in" for the further care and preparation needed for exhibition. At this age cockerels must be divided from pullets, and the chicken period may be considered over.

CHILLED EGGS.

The moment at which an egg is fatally chilled is not certain. If a hen is off her nest for twenty minutes in cold February weather the eggs will be chilled, but not to death. Endeavor to get the hen on in ten minutes on a frosty day, or cover the eggs with a layer of cotton-wool, flannel, or even hay. Some hens *will not* be hurried, and to try it will only bring trouble. A shorter period of *chilling* will destroy vitality in eggs during the first stages of incubation than a longer period when the chicks are nearer perfection. The first thing to do if the eggs are chilled is not at once to force the hen on to her nest, but to immerse the eggs in warm water at 105° or even 107°. Meantime get the hen back on to some false eggs; and when the chilled ones are thoroughly warmed through, replace them under the mother. Valuable eggs should not be despaired of even if the hen has been off for some hours, but should be treated as above, and next day, if examined with aid of the eggtester their vitality, if it is not destroyed, will be clearly seen.

CHOICE BETWEEN DEFECTS.

In breeding-pens it is at times necessary and politic to breed from a bird which may have a marked defect as a show-bird. If unable to procure perfection, choose a cock which is quite free from that particular defect from which the hens may be suffering. If, *i. e.*, the hens black Hamburghs or Andalusians have red splashes on their ear-lobes or white on their faces, choose a cockerel of very good strain and entirely free from these evils to mate with them, even though his color, his tail, or his comb be *imperfect*—rather than one perfect in all these latter points and failing in the former. If the hens are meagre in comb, mate with a male bird carrying a comb large to a fault. If the cockerels' combs are coarse and large, mate with hens whose combs are extra small, and so on, making use of defects to produce perfection.

CHOLERA.

Caused by want of fresh water and green food, and means of sheltering from excessive sun. The symptoms are those of aggravated diarrhœa, great weakness, and constant thirst. A tea-spoonful of caster oil with five drops of laudanum should be given, or 5 grs. of Gregory's powder, in place of the castor-oil, with the same dose of laudanum, every five hours till the diarrhœa ceases. Soft food with cayenne pepper, fresh green food, and as little water as possible. The bird must be sheltered from the sun, also from damp. Such severe diseases should not arise, they come from mismanagement and neglect.

COCKERELS.

May at a very early age be distinguished from pullets by the incipient spur, the more developed comb, and tail-cushion. At from eight weeks to twelve remove the cockerels from the pullets, and keep them separate for six months at least. At four months weed out into the fattening pens or yard all those birds which do not promise for exhibition or prize stock, and treat the few remaining accordingly. Any very promising bird, if observed to be shy and worried by more advanced or pugnacious cockerels, should be given a run to himself with a few companions of whom he is not afraid. A bully in the cockerel pen is a source of much mischief, as he prevents valuable birds from feeding, and thus getting on well. Do not spare bone-dust, lime, oyster-shell, green food, and meat, and unlimited range to growing birds so as to get the frame and bone well developed; the fat will be easily put on afterwards. Sound grain of wheat, barley, buckwheat, or Scotch groats—as much as will be picked up *clean*—should be given, and for soft food table scraps, Spratt's food. oat and barley meal; these will keep birds in good health, glossy and firm in plumage.

COCKS.

Should not be kept over their third year for breeding unless of especial value. Early in the season adult cocks are not lively, and the eggs from their pens are not as a rule fertile.

They should be mated with pullets, not with adult hens. During moult remove the cocks from the hens. The former can then live together, and will not fight if they have free range, but care must be taken that all food is well scattered, or placed in many pans, that all should feed well. To insure fertile eggs, cocks must be fed apart from their hens when mated in December, till March, as their gallantry often causes them to go on short commons. When once the breeding pens are made up, do not change the cocks about; they become fond of the hens, and it is a great risk disturbing the family during the hatching season. Some cocks are very sulky and even savage to the hens during moulting time; if this is perceived, at once remove the bird and give him a pen to himself; he requires rest and good stimulating food, and with new plumage good temper will return.

COLDS.

Generally caught by birds kept in damp and ill-ventilated houses. First symptoms must be taken in hand at once, as an ordinary cold neglected may turn to roup. The first sign is sneezing and running at the nose. Isolate the bird at once, place it in a warm draughtless place, wash the nostrils with vinegar and water, give a tablespoonful of castor-oil to a very large bird, less to a smaller, feed on soft food, oatmeal warmed in beer, or bread and beer, with plenty of grass. An excellent thing to stop the running at the nose is to fill the nostril with a pinch of snuff. Wash the face well in warm strong tea, and put the fowl in an exhibition basket, out of draughts. A pinch of Epsom salts may also help. Colds are *most* catching.

COLD WEATHER.

Frosty cold is less dangerous than damp; to insure success in poultry keeping, keep houses and covered runs dry. Let the weather be ever so frosty or wet the birds will not suffer if allowed free run in the day for some hours, provided home shelter be perfectly dry and free from draught. Breeds with excessive comb development, *i, e.*, Minorcas, Leghorns, &c., during hard weather should have lots of straw about in the covered run, and mats should be nailed up to shelter from keen east winds. Do not allow birds to drink snow water, and it is well to give prize stock warm water in their tins in severe frosts, it keeps unfrozen longer; and is more comforting. It saves much trouble to empty out all water vessels in severe frost at night. Soft food should of course be given warm in winter, and the grain at night may with advantage be warmed through in the oven *dry*, without water; the birds relish it, and it is a source of warmth and comfort through the long winter night.

Hampers for exhibition birds should have an extra swathe of scrubbing flannel put round them in severe cold weather, allowing, however, ventilation at the top of the hamper, or the birds will arrive at their destination with *blue* combs.

Vaseline rubbed on the combs of breeds subject to frost-bite is an excellent preventive. The food should have a dash of red chillies, black pepper, or some of the many excellent condiments advertised. Some stimulant in severe weather may be used to advantage.

COMBS, THEIR VARIETIES.

Double or Rose Comb, as in White Dorkings, Black Hamburghs, etc., etc., a flat square comb, wide in front and narrowing to a peak, pointing backwards, evenly serrated, presenting an even surface free from hollows or elevations. Pea Comb, as in Brahmas, a triple or ridged comb, the middle higher, longer, and thinner than the side ridges, the whole being firm, of moderate size, and firm on the head. Single and pendent Comb, as in Spanish, Minorcas, Andalusians, Leghorns, etc., etc. The single comb is upright, quite straight, free from any excrescences, fine in texture, having no thumb-marks, and evenly and regularly serrated. Pendant combs are seen in hens only of the last-named breeds; fine smooth, evenly serrated, without side sprigs, they must fall gracefully over the side of the head and face, taking a slight curve to one side before falling over on the other.

COMMON FOWLS.

If a paying poultry-yard is desired, do not run into the error of purchasing a "few common fowls;" and if poultry are kept "for pleasure," this will not be attained with barndoor mongrels. Should "a few common fowls for eggs" be desired, a cock and five pullets of Andalusian, Minorca, or Hamburgh breed are the best, and such birds can be purchased young in July from any of the well-known breeders, pure and true to breed, but set aside by them as not perfect enough for prize-stock or exhibition. If time allows, get a first-class sitting of eggs, and hatch them in March; you will then have a pen giving pleasure to the eye, and more paying too than any barndoor deteriorated cross-breeds. A cheap coop for the common cock and five pullets can be made out of a very large sugar hogshead, whitewashed inside, with split pine-tree top inserted across for a perch, and a pot basket hung at the side on two strong hooks for the laying nest. Cover the cask with tarred felt, coming over the top and down the sides; it will give dryness and warmth. If expense is no object a dry outer shelter is to be desired, and free range.

CONCRETE.

Nothing is so good for the flooring of all houses as the original soil over which they are built, dug up and mixed with screened mortar rubble. If rats abound, it is advisable to concrete the floors of roosting places, but they must then be covered deep in screened dusting material of some sort. Birds kept on concrete flooring without these precautions, however well swept and cleaned, will not flourish.

CONDITION TO KEEP BIRDS IN.

This depends on constant attention to a thousand small details: proper diet, clean water, and perfect cleanliness, of course; abundant sand and mortar, always pure and dry; grass runs, which must be kept mown if for feathered-legged Asiatics; perches arranged so that the cocks' tails do not rub and get injured against the walls, entry trap-doors not too low or narrow; covered runs available so that in wet weather the birds have light, air, and space to scratch amongst dust and straw without roaming out in the wet, and shelter for some breeds from the scorching midday sun; water-tins arranged so that they cannot upset; paint, tar, and large water-tubs kept out of reach; ducks kept out of the poultry runs; fights and sparring through wire partitions prevented; cleaning of the poultry-houses done first thing every morning, and not left till the birds have trampled the droppings about all over the place. If birds are to be caught for any purpose, or counted, it must be done at night to save beating about. If necessarily done in daylight, a piece of dark cloth drawn over the window will enable the poultry-man to catch the birds without injury or fright. When whitewashing, put grease in the wash, or it will be likely to come off and soil the plumage. Laying boxes, or hampers, and nests of all kinds should be constantly replenished with clean short straw, and should never be so small that the tail and breast feathers rub as the hen turns about. All partitions and wire-work to be kept free of jagged crooks or points. Injured feathers cannot be replaced till yearly moult.

BEST BREEDS FOR CONFINEMENT.

Decidedly Andalusians and Minorcas. They can be kept in high condition, lay splendidly, produce fertile eggs, and win prizes in the most confined space, all conditions of health being attended to; but the smaller the space the greater must be the labor and care to be expended upon it to produce any great results. Leghorns are good, but their spirits are more depressed than the Andalusians by penning up. Brahmas and Cochins of all classes will be content to squat all their lives in a pen a few feet square, but they are apt to grow fat and diseased, and lose all gloss; and when not eating they are brooding, and neither occupation fills the egg-basket; they are, however, most tractable, and nothing looks handsomer then a pen of Light Brahmas confined in a villa-garden by wire two feet high strained round a portion of lawn-grass: it will effectually keep within bounds these models of contented domesticity. Andalusians and Minorcas require wire-fencing, without woodwork at the top to rest their feet on, at least seven feet high.

COOKING FOR POULTRY.

A little trouble in this respect will be amply repaid in the poultry-yard. Every establishment where 100 head of poultry are kept should have its lock-up food store-room, and if a gas-stove can be put up its help is invaluable. House-scraps can be regularly brought out to the food-house hot from the kitchen by 8 a. m., and with boiling (not cold) water let meals of all sorts in turn be mixed with the scraps till it forms a crumbling mass. All food for ducklings is better given warm than cold, chickens also appreciate their milk and their porridge with the chill off. Liver given raw is not palatable, but if put in water over the gas-stove for ten minutes, and chopped, hot with Spratt, and thrown to the birds in pellets, it is greedily devoured, and more good is got out of it. Grain baked in the oven dry, and given warm to the birds, is very good in the winter-time.

CORN.

The best staple foods are, 1st, buckwheat; 2nd, wheat; 3rd, barley. Hemp may be given occasionally; and as for Indian corn, it is better to give none than to give constantly, it is much too fattening for laying or breeding stock. It is of great advantage to give a change. Buckwheat promotes laying, aud is also good if birds are poor and out of condition; for young birds it is excellent, and being smaller and more fattening, it comes before barley. Hemp must not be given regularly, being to heating; if persevered with it will cause loss of feathers, but it is a necessary help in preparing birds for shows. Do not be persuaded to buy "sweepings," or "poultry mixtures," or "poultry grain;" these preparations are simply injured grain of all sorts mixed up together, some a little and some very seriously damaged, so as to be unfit for food, but in the bulk it is hard to find this out. It has another disadvantage: you cannot give your fowls change of grain, which is most beneficial. Sound, good grain is always economical in the end, care being taken not to leave it on the ground, but to feed in moderation, so that not a grain is lost. Oats are not much relished by poultry, having too much husk; they are altogether the least profitable grain, though cheapest.

CRAMMING.

Ordinary cramming or fattening such as it is possible for the amateur to carry out is simple. In a shady and cool outhouse or shed, the stall of a disused stable or cow-house, or a poultry-house, confine the birds for fattening, six or eight, according to the size of the enclosure. Sand the floor deeply, keep clean, give little light, and feed with soft food three times daily, with as much as the birds will consume without leaving any behind. In this case liquid food cannot be given, though highly thought of. Buckwheat, maize, oatmeal, and barley-meal are all excellent mixed with half-and-half milk and water, not too dry. Birds for this purpose should be about four months old, and they should be ready in a month at the outside. Perfect quiet and cleanliness are needful, and as soon as fattening is completed they should after twelve hours' fast, be killed. If kept too long after the process is complete, emaciation and sickness set in, and the work is undone. The French process of cramming by machinery is too complex and lengthy to enter upon here, and any one intending to carry it out should study some of the larger works on poultry.

Another simple mode of cramming consists of confining the birds each in a coop, with floor of round bars, through which the droppings fall and can be cleared away daily; three times a day take them out and force pellets of milk-mixed meal dipped in water till the crop is full. If digestion has not cleared the crop out, a meal must be missed. Do not confine the birds where they can see or hear their fellows. Perfect coolness, quiet, and shade to semi-darkness is needful.

CRAMP.

No. 1. If in the legs, foment or stand the bird in hot water, bandage with flannel strips half an inch or one inch wide, according to size, and, put it in a dry, warm greenhouse, with a sandy, gravel bed, and straw-chaff litter. Give tonics in the water.

No. 2. Rub the legs with hot oil, turpentine, or camphor.

No. 3. Wrap in flannel, and keep by a fire at night; give free dry run by day, diet ample and nutritious, with meat. For a chick three months old, opium (quarter grain) three times daily is a help.

SITTING HEN CRAMPED.

Rheumatic or gouty hens only are subject to this; they cannot stand when they get off the nest. Give four drops of sal-volatile occasionally; rub legs with turpentine and oil daily.

CROOKED BREASTS.

The usual theory as to crooked breast-bones is that they are caused by giving perches to chickens *too early* in life, or by the perch being too large or too narrow. Experience shows, however, that crooked breast-bones are found if the strain is a weakly one and wanting in constitutional vigor, or if the birds are pampered and brought up at high pressure, in confined and over-crowded pens, where stimulating condiments and excess of feeding takes the place of free range, fresh air, and ample, nutritious food. White Dorkings brought up with and treated in all respects like their companions, Andalusians, suffered (under my care) with this weakness, while the Andalusians, treble their number, showed not a single example of this evil. High perches in a small roosting-house, where birds have no space to make a sweep which they would in freedom take from the highest tree without injury, are highly mischievous, and cannot be condemned too strongly; but these, though they injure the breasts and ruin the birds, do not account for the peculiar bend in the ordinary "crooked breast-bone." Birds so formed should not be bred from, for prize stock.

CROP-BOUND.

One of the most common of crop ailments, caused by improper and excessive feeding, by lack of gravel, mortar, lime, and gritty rubbish, which birds eat so greedily, and which are necessary for digestion. The crop becomes choked up and distended with a solid mass, often very hard; the bird cannot eat, mopes about, and loses flesh. Pour warm water down the throat if a slight case, and work the mass about with your hand, starve the bird and give no water. A teaspoonful of gin now and again is good;

when softened give castor oil, a dessert-spoonful to an adult. After recovery feed very little for a few days. If these measures fail cut the crop open at upper part, and with the handle of a teaspoon remove the contents, oil the finger, and feel that the outlet to the stomach is free, then sew the parts up, the crop and the skin being sewn separately. Horsehair is better than silk, which does well, however. The bird will soon want to feed ravenously; give bread and milk, not too wet, and keep the bird quiet for a few days, on short commons of sort food.

CROP SOFT OR SWELLED.

Contents feel like liquid, and the crop sways about as the bird walks; this does not affect health or spirits. Puncture the crop and let the fluid out, feed very slightly, with soft food, and hardly any water. The worst months for these diseases are August and September or October. When wheat newly thrashed is given to poultry they eat greedily, and as new wheat swells in the crop, the whole mass hardens, and closes up the outlet into stomach, bringing on a disease which must prove fatal unless properly treated. Exhibition birds kept for a week in a pen eighteen inches square and overfed on whole maize, often return with the crop overcharged; soft food only should be given if this is the case, and a teaspoonful of gin, brandy, or port wine, to stimulate digestion.

CROWDING POULTRY.

One of the commonest of evils, and fatal to success. Most amateurs go in for several breeds of poultry; this, unless a park or farm is available, is very unwise. The birds may be kept in comparative comfort during the winter months, but in the breeding season, when the chickens begin to come, and in August, when pullets have to be separated from cockerels, and these again in October kept separate from adult hens—when these also have to be parted from their mates, and exhibition birds require each their roomy and separate pen, it is impossible to rear many breeds successfully, each having its perfect exhibition specimens; for this, space is a matter of necessity.

DEFORMITIES.

Squirrel Tail.—This is a common fault, and difficult to breed out. The tail projects in front of a line drawn perpendicularly to the end of the back, sloping, in a marked way, towards the neck.

Slipped Wing, or Turned Wing.—An accident to which some breeds are peculiarly liable when making their adult plumage. The primary feathers, instead of being tucked in when the wing is closed, protrude; appear twisted outwards and in disorder, the inside of the feather coming outside. Where this is the case, a cockerel has little chance in any good competition; it is incurable, and, moreover, hereditary. When the flight feathers only hang down, as if too weak to be compactly tucked up, but not otherwise disordered, cure is possible when taken in time. As soon as displacement is seen, tuck the wing up every night at roosting-time, and when the bird is more advanced the wing should be bound near the shoulder as tight as possible. From the outside centre of this ligature a cord must be passed round the shoulder, and fastened to the inside centre to prevent its slipping off. Stout whipcord may be used, or tape, and every care taken to have each feather in exact position before tying up. From six weeks to two months will be required to effect the cure, and even expert hands have difficulty in putting on the bandage satisfactorily. Night is the best time to operate, for the sake of quietness.

Vulture Hock.—This is caused by the feathers projecting considerably beyond the hock joint in a stiff, awkward manner.

Asiatic breeds are particularly subject to it, and if a bird has this deformity strongly marked, it would be useless for breeding or for exhibition purposes.

Wry tail.—A sign of hereditary weakness; occasionally comes from a bird being kept in a small pen, carrying its tail habitually on one side as it walks round. If a habit only, it may be cured by snipping out a portion of skin on the side from which the tail inclines, and the scar in contracting draws the tail back to its proper position.

DIARRHŒA.

Caused usually by cold and wet weather; in chickens, often due to improper diet. First try boiled rice mixed with powdered chalk. If the case is severe, which it should not be in a well-ordered yard, try any of the following:—Two or three drops of cholorodyne in water occasionally, half a teaspoonful of brandy twice daily in water. A pill containing five grains of chalk, five grains of rhubarb, three grains of cayenne, half a grain of opium. For food, rice boiled in milk, only, if the case is bad. Barley is the first grain used on recovery for hens, pearl barley for chickens. *Bone-dust* is an excellent preventative; young chickens should always have it. (*See* BONE-DUST.) Ample green food should be at hand.

DIRT AND DROPPINGS.

These left on the ground or under the roosting-places, are most prejudicial to health. They should be collected daily, and not swept or raked about in clearing them away, so as to become mixed up with the soil, but carefully shovelled up separately. It is advisable to have a large covered box or tub, into which you can throw the droppings, without other rubbish. This manure, if kept dry, is equal to guano, invaluable for garden and farm land.

The produce of vegetables and strawberries, when it is judiciously used, is trebled. If earth or ashes are sprinkled over the manure, there will be no unpleasant smell. Fifty hens are said to produce in the roosting place about ten cwt. per annum. It is best to use it at home when possible, as gardeners object to pay for the rubbish which is mixed with it. In large towns leather-dressers will give a low price for it.

DRAINAGE.

Where the ground is wet this is very important; damp is fatal to poultry. A proper fall should be arranged in the ground, and a small expenditure on drain pipes and the making of dry wells will be amply repaid by freedom from disease. Drainage from roofs may be very useful if collected in a water-tub with a tap for washing out feeding tins and water fountains. If the roofing be made of any material to which tar is applied, care must be taken never to use the water for drinking purposes; it is most unwholesome for poultry.

DRINK.

The feeding of poultry, though too often done in a senseless manner, is, perhaps, not so grossly neglected as their water supply. Many poultry fanciers seem to think there is no occasion for a regular, still less for a clean, supply. When water-tins are refilled the remaining sediment of dirt is too often left to taint the fresh supply. Birds are left for long hours without water, then suplied in excess and allowed to gorge themselves, much to their hurt. Properly, water should be supplied fresh twice in twenty-four hours; it should be fresh spring water, not rain and the vessels should be emptied and rinsed well out in a bucket. A green slime often coats vessels in which water is continually standing; this should be scrubbed off daily. The attendant should go round with a bucket of hot water in which all fountains should be scrubbed, and care must be taken to stand water out of the sun, as the heat renders it unwholesome. On return from a show, give but a very little water on first returning to the home pen; if taken in large quantities it is apt to turn the comb to a dull blue color. A regular and pure supply of water is more important than is generally supposed. If ducks are kept means must be taken to prevent their polluting the fountains by washing in them, and if standing in a crowded pen it a is good plan to place the water vessels on a stand, or to have tins supplied with two hooks by which they can be hung up on the wire partitions of the pens, out of the reach of the dust and droppings which will otherwise be scratched into them. Snow water is injurious, and rain also if out of tubs drained off a tarred roof. Water in which potato parings have been boiled is injurious, and should not be used for mixing meals. A good water vessel for hot weather is an earthenware saucer with a slate over it, in which a hole has been cut to allow the bird's head to go in. A flower-pot with a cork in the drain hole, filled with water and then reversed in a saucer, is an excellent water vessel, keeping the water cool by refrigeration. Tonics in water are a great help in the raising of prize poultry. Fattening ducklings must have very little water to drink. Adult fowls should have it at discretion. Whether chickens should have a constant supply is a moot point. Some declare that they never give any, and that the chicks are extra healthy, others allow it to be always at hand. Chickens running with their mothers in a state of nature drink when they come to a puddle, apparently without much attention to rule or theory; but those hatched in artificial rearers are certainly somewhat too much addicted to drinking, and as long as water is supplied they will go on imbibing. The best plan with these is to give milk once a day (in the forenoon), a reasonable allowance, and to remove it when

HOUDANS.

all have had some. Water midday, and water with tonic at night, removing these also when the thirst is quenched.

DROPPING EGGS.

This is caused by a too stimulating diet, also by want of mortar, oyster-shells, or grit for shell formation, also by the hens being too fat Feed less, give no meat for a time, vary the diet with rice, potatoes, sharps, dari, and wheat. Give a dose of castor-oil, and iron tonic in the water. Should this not cure the evil, give one gr. calomel, one-twelfth gr. of tartar emetic.

DUBBING.

This operation consists in the removal with a sharp knife of the comb, ear-lobes, and wattles in game birds. In the days of cock-fighting this was done to save the combs of the fighting birds from injury, and to remove what afforded a good hold for the enemy. Now it is merely a fancy point for the show pen. Cocks should be dubbed when getting their adult plumage and when their combs are well formed. Game and Game Bantams are the only breeds on which this cruel operation is now practised.

EAR-LOBES.

Sometimes called "deaf-ear," a more or less pendent ornament to the face, just below the real ear, varying from pure kid-like white to red and blue.

EARLY OPENING OF HOUSES.

This has much to do with health, and if birds who rise with the sun are shut up in close, ill-ventilated roosting-places till 7 and 8 a. m., no success will attend the mismanaging owner. The roosting-house should open into a covered run which the birds can enter at their own free will, to find a little food and to amuse themselves till the attendant comes his rounds, which he must do in summer at 6 a. m.

EARLY-ROOSTING.

Chickens of some breeds even two and three weeks old are very fond of roosting. Care should be taken to remove anything high, for the danger is great to the breast-bone if they come down from a height when very young. But it is well to provide roosting-places, such as reversed boxes, covered with sand or straw, or broad perches very near the ground, as the exercise and amusement of hopping up and down is good, and if placed in the sun, the chicks will delight to congregate upon them and preen themselves after meals. For night roosts perches are not desirable; shelves sanded and covered with chopped straw are best. Under any circumstances, perches should be broad, so that fowls can conveniently stand up and walk along their length without any difficulty in balancing themselves, and not too high.

EARTH DEODORISER.

A supply of sifted earth and sand should be at hand (where the birds cannot taint it) for the purpose of deodorising and re-earthing the floors of all artificial mothers and sleeping shelves. Nothing destroys offensive smell better than earth, and if freely used, with occasional dredgings of sulphur, bad air and vermin will be avoided. The earth should be sifted or screened. If small chicks sleep on lumpy beds of earth, the delicate and fragile breast-bone may be injured by the pressure of even one night's lodging on an improperly made bed.

ECONOMY

In food may be secured by buying the best uninjured grain, and seeing that not more than one woman's handful of grain is given to each fowl for a meal; this is enough, and none will remain on the ground. In building houses it is an economy to remember that at timber yards wood is sold cheaper in short lengths than in long, and that if the length for partitions, etc., be fixed on at first a saving is effected by ordering all to be sawn to lengths at the timber yard; by this labor and material are economised. The price of boarding also varies as much as does the silk for a lady's dress. Accurate measurements should be made, so as to avoid cutting up to waste.

EGGS.

Addled are those which, being unfertile, get jarred or chilled, and are then sat upon to the end of the incubation period, and become decomposed. They exude an offensive moisture, and if broken, explode with noise and smell. Remove such eggs as soon as detected, as their presence in the incubator or nest is fatal to success in hatching.

Barren or Unfertile Eggs may be taken out of the nest a week or five days after they have been sat upon. They should be examined with an ovoscope by the aid of a candle or lamp; those which are barren will be found perfectly devoid of veining, and the yolk will be seen to sway about from side to side with the least movement of the hand; such eggs are fit for cooking, and perfectly wholesome.

Blood-stained Eggs.—The first laid by a pullet or hen after moulting may be slightly so, and it need not give any cause for alarm.

Broken Eggs in the Nest.—Plunge those remaining unbroken into warm water at 105°, and wash them clean while the straw of the nest is replaced and all made sweet. If any of the eggs are only cracked, at once apply sticking-plaster or gold-beater's skin, and patch the crack up firmly to exclude air. The chicks will hatch out in most cases quite well.

Chilled in Hatching.—The time at which the chill becomes fatal seems to vary according to circumstances, weather, position, period of incubation, and so on. Twenty minutes exposure in cold weather, and in a draughty place, will addle a sitting, and yet chicks have come out when the hen has been absent an hour or two and even more, and the eggs have been quite cold. Eggs are more liable to chill after two or three days' incubation than towards the close. If a hen leaves her nest, and the eggs are found chilled, place them at once in water at 105° while you secure and settle or replace the hen, and let her sit on to the end of her time; the ovoscope will here be useful, as it will enable you to see if life still exists, and if not time may be saved by resetting the hen.

Color of Ducks' Eggs.—Varies from pure white to cream color and to bright green. Pekin and Aylesbury ducks are no exception to this rule; it does not affect the produce in any way, nor can it be accounted for.

Color of Hens' Eggs.—Asiatic breeds lay eggs from deep chocolate through every shade of coffee color, while the Spanish, Hamburg, and Italian breeds are known for the pure white of the eggshell. A cross, however remote, with Asiatics will cause even the last-named breeds to lay an egg slightly tinted.

Egg-bound.—The best cure for this is a table-spoonful of warm treacle, into which chopped groundsel is mixed, giving it warm.

Fertility of Eggs depends on the number of hens to each cock, on the space allowed for the run, if any, and on the age of the male birds; all efforts to ensure it are negatived by improper and over feeding. The eggs of very large exhibition birds, bred for size and feather, are very often unfertile, and eggs from pens where the birds are constantly travelling to shows are, as a rule, less fertile than others. Leghorns seem to suffer less from this than some other breeds. Changing cocks about in the midst of the breeding season will make the eggs unfertile, though it may be only to exchange them from one home pen to another.

Foretelling Sex of Chicks.—No rule can be laid down about this, and the shape of the egg has nothing to do with it. Early broods bring most *cocks*, late broods the pullets. This is generally the case, but no rule is reliable.

To Hatch Early.—An incubator is a great advantage for this purpose: in November and December, January and February, sitting hens are hardly to be got, and the incubator can be worked independently of weather. Chills must be guarded against, and are specially dangerous early, the fertility of the eggs being less vigorous than it is later on, and a batch of eggs may be easily addled. Brahma hens kept in a warm place are the most likely to sit early. Nests should be made in a sheltered warm situation, as excessive cold will cause a hen to leave her nest; and if she is comfortably housed and her food given warm (baked grain dry) in severe weather she will be more likely to keep steady. While the hen is off feeding pour boiling water round the nest daily during the third week. In severe frost a handful of dry hay cast over the eggs while the hen feeds keeps them from to severe a chill.

Increasing Egg-production.—Mark those hens in your flock remarkable for the size or the number of their eggs, and hatch their eggs in preference for laying stock. Choose breeds which do not sit. Do not over-feed or fatten, and keep laying hens in an active hungry state; do not, however, run into extremes and under-feed them; they must have plenty and yet always be ready for food. Do not keep old hens, two years is the *outside* limit. Birds hatched, say, in March, 1884 should, on an egg-farm, be killed for table on the first signs of moult in autumn, 1885; they are then quite young and fetch a good price, and will *not* be so valuable in 1886. Laying hens should not have to much fattening soft food, sound grain in variety is the best diet, and plenty of green food, oyster-shell, and mortar rubbish.

Preserving Eggs for Sitting.—If the eggs are from prize stock, each egg should be marked with the date or number of the prize pen; the mark should be placed on the small end of the egg, otherwise the chick in pecking out is apt to obliterate all traces of it. The eggs should be collected in a basket containing bran or chaff, and afterwards placed in a box (with a lid,) in bran, *small end up;* they should be entirely covered up in the bran. Perfectly new-laid eggs *must* be used for incubator work, under hens staler eggs will hatch. I have known eggs hatch when placed under the hen at fourteen days old, but new-laid eggs are always to be preferred.

Moistening Eggs.—In hot dry summer weather watering the hatching nest with water at 105° from a fine rose water-pot while the hen eats her daily meal is a good practice; in cold weather pour boiling water round the nest. In an incubator do *not* water the eggs, but get the moisture required from wet earth trays.

Preserving Eggs for Winter Use.—To ensure success, whatever the keeping medium, place the eggs into it as fast as collected from the nest. If they lie about here and there, exposed to air and sun and movement even for a day, the result will not be satisfactory. The ways to preserve eggs for winter are various:—

No. 1. Dissolve quicklime in water, and add a little cream of tartar, put in as laid, and see that the water quite covers the eggs.

No. 2. Rub the eggs with lard, or butter, or oil, and immerse in bran.

No. 3. Bury the eggs in powdered unslaked lime.

No. 4. Bed in salt.

No. 5. The French method: Varnish the eggs with varnish of linseed-oil, and beeswax.

No. 6. Smear the eggs over with linseed-oil, and place them in dry sand.

No. 7. Rub in butter, and store the eggs in boxes well closed down, pasting paper over the cracks so as to exclude air: keep in a cold place.

Round and Long Eggs.—Round eggs are said to produce pullets; long eggs cockerels—a fallacy, and proved to be incorrect.

To secure Eggs all year round.—Keep an incubator going; and if your runs are on a small scale, hatch out, say twelve chickens every month, beginning in December and continue till May. These should provide pullets enough to keep the egg-basket full all the year round for a small family.

Selecting Eggs for Sitting.—*Very* round and *very* pointed eggs should be rejected, also very small and extra large ones. Double eggs never hatch well, they produce deformities. Medium-sized eggs are more vigorously fertile than others. Eggs that are rough-shelled or brittle are bad, and crooked, misshapen eggs bespeak a delicate, consumptive strain of fowls. Never place dirty eggs in a nest: wash in tepid water with a sponge till clean. Ducks' eggs particularly, should always be cleansed. In an incubator, where so much depends upon the purity of the air, this is doubly important. Water for washing should be 105° F.

Soft Eggs.—Hens often lay soft eggs, devoid of shell. This arises from several causes, the most common being the want of material to form shell where birds have only earth or grass or a paved yard. Another cause is over-feeding with rich, soft, stimulating diet. Reduce the quality and quantity; give more dry dari and wheat grain, less of soft food, and add powdered oyster-shells, mortar-gravel, grit, or the cleaning and chippings of lime from boilers. Nests dirty and insufficient in number will cause hens to drop their eggs prematurely; the egg gets broken, and then not unfrequently the juicy morsel is eaten up, and a taste for egg-eating is formed. If proper management does not soon avert or cure the evil, try physic:

No. 1. Diet on rice and potatoes sparingly given, and a pill of one grain of calomel, with one-twelfth of grain of tartar emetic.

No. 2. Dose well with castor oil; put old nails or old keys in the water-tin, or two lumps (size of filberts) of sulphate of iron to the gallon of water.

Test of Fresh Eggs.—The fresher the egg, the smaller the air-chamber. This can be seen at the broad end of the egg if it be held up against a strong light in a dark room. Stale eggs have a mottled, greyish look about them; and a new-laid egg will always give a feeling of warmth if the tongue is pressed to the large end.

Small Yolkless Eggs.—This, again, is a sign of over-feeding—and most likely an exclusive diet of Indian corn, with little exercise and no green food.

Egg-eaters.—Such hens must be sharply looked after. Where only a few fowls are kept near the house, the evil has been overcome by taking the eggs as soon as laid, but this is impossible in large runs; and unless very valuable for prize breeding, the offender should be killed at once; it is a fault easily taught to other hens. A good plan for finding out the culprit is to put some eggs about the yard, or run, and to let the hens out; the offender will rush at the egg and commence breaking it, and in a moment will swallow it down. Next, place some stone eggs about, and one real egg, *blown*, and filled with mustard and cayenne pepper; the offender will peck away at the stone eggs, and at last seize on the hot one. This lesson is sometimes, but not always, successful. Nests adapted to save the eggs of egg-eating hens are advertised in the poultry-journals, and are useful should it be necessary to keep birds which have contracted this very objectionable habit.

EPSOM SALTS.

A small quantity given occasionally, mixed into the soft food, encourages laying, but care must be taken not to give much, as it will produce scouring; one tablespoonful to 25 birds is enough.

ERUPTIONS.

Scurfy heads, or scurfy scabs on the ear-lobes. This state of things indicates want of green food, and that the birds are kept in ill-managed dirty yards. Castor oil two or three times at intervals of two to three days, and powdered sulphur in the food. Turmeric and cocoanut-oil ointment may be rubbed on the scurfy parts, and all will soon be well.

EXCESSIVE FEEDING.

One of the most common causes of disease and loss in the poultry-yard. No food should ever be left on the ground; any so left shows that the birds are getting more than is good for them. A good plan for the inexperienced poultry-feeder to go by is this: At the morning meal, which is of soft food, place it in pans or tins, several, according to the number of the fowls, so that all may have a fair chance of feeding without overcrowding; leave the birds for twenty minutes, after which return and remove any remnants left, leaving none behind. The birds will have eaten as much as they require, and over that mark there is danger of bringing on internal disorders of the liver and stomach, loss of feathers, laying of soft eggs, unfertile eggs, and every poultry ailment under the sun.

FAILURES IN POULTRY KEEPING.

Chiefly due to a mistaken idea that poultry can be made profitable and a source of wealth or income without any outlay, everything being expected from whilst nothing is done for the birds. Some of the most common causes of failure seem to be from want of cleanliness, over-feeding, and that on injudiciously arranged diet, from trying to keep too large a stock of fowls for the space available, and, perhaps, too many varieties. Want of dry shelter is a common cause of disease. It is not enough for the roosting-house to be dry; there must be a space where air and light can enter freely, where the birds will resort on wet days to dust and preen themselves. If poultry is kept for sale of eggs and chickens, etc., a town or city near at hand is indispensable; and the soil should be dry and sandy rather than clay, with plenty of shelter in the way of orchards or shrubberies. Want of knowledge as to trussing and preparing poultry for market is one cause of the low prices paid to amateurs for their produce, and another drawback is that so many start poultry-work with old barn-door birds, of very inferior type, and poor both as layers and fatteners.

FANCY POINTS.

It is a usual remark that fanciers "breeding for feather" or for "fancy points" are really doing harm, inasmuch as the breeds deteriorate in their hands for all economic purposes; and yet it is very certain that the improvement in useful farm poultry of late years is due to the fancier, for it is from his runs that birds kept for commercial purposes only are crossed and resuscitated. Birds kept for table purposes, and as egg-producers, are seldom of pure breed; the first cross is sought for, and it is out of the hundreds raised by the breeders of pure stains that the crosses are obtained, and the race improved for economic purposes. Had the fancier no high standard of feather, etc., to breed to, and that difficult of attainment, there would be no object in his hatching and feeding with such scrupulous care the hundreds of pure birds he now produces yearly. To enable him to pick out three or four, or even one, ideal bird, true to fancy points and fit to win at large shows, some hundreds must be hatched and fed extra well from the first, and it is those that fail of perfection which give the useful first cross for economic purposes.

FARM POULTRY.

The present system of keeping farm poultry cannot answer, but great strides are being made towards an improved state of things, and the subject is being thoroughly ventilated. There is no doubt that on a farm there are special advantages for poultry-keeping on a large scale. Barns and extensive sheds for sheltering carts, etc., are at hand, which, without building especially for the fowls, can be adpated to supply admirable poultry accommodation. The ricks of hay and wheat, etc., if properly raised from the ground to avoid harboring rats and damp, also give shelter; and even if the farm is small it should be capable of supplying the greater part of the corn and meal used, for birds, with ample range and the gleanings of the farm-yard at command, forage largely on their own account. System and attention are wanted, and common sense, which the woman of the farm, to whose management the poultry are left, as a rule, do not possess. Old birds are kept year after year, from which supply of eggs must be small, and, as a rule, the hens are fattened and the chickens under fed. No pains are taken to hatch early, and thus eggs are not to be had during those months when they gain the highest price. Feeding is irregular, and the food given is seldom varied, whereas we know that variety is specially good for poultry. The farmer's wife seems unable to name any other grains but Indian corn and wheat; whereas she should grow, besides these, a small patch of barley, buckwheat, and sunflowers, also oats for grinding to meal. Fresh water is never supplied, the birds usually drinking from the drainings of the manure-heap, and the poultry roosting-house is cleaned out, perhaps twice a year; the manure is supposed "to keep the fowls warm." The number of birds too, as compared with the acres at command of grass, wheat, turnip, and other fields, is absurdly small; and whereas the fancier with love for poultry will within the space of seventy feet by fifty produce and rear over 200 birds for table, egg, and show purposes, the farm poultry on some seventy acres will not count over 100 head, if so much. The farmer will carefully breed from his best stock, the best milkers, and the best fatteners, in cattle and pigs, the sheep producing most lambs, all are carefully perpetuated, but no trouble is taken about the progeny of the best laying hen. Mongrels are started with, and breed in-and-in year after year; the race deteriorates in egg-production and in size; whereas a pure breed, or at least a first cross, should be started with, and these birds should be mated with a pure-bred cock of another breed—fresh blood should be introduced from time to time. With all his blind neglect and ignorant mismanagement going on, it is no wonder that farm poultry pay pennies only where they should bring dollars. Many farmers consider that poultry injure the crops. It must be said that some thirty or forty heads roaming over a field do forage and trample on pretty freely; but the question is whether the amount of good they do in killing insects of all kinds, especially slugs and wire-worms, does not compensate for the leaves they consume and destroy. It is the opinion of many high authorities that the good they do in this way exceeds the evil.

FARMYARD DUCK.

Here again, the farmer as a rule is content with the variest mongrels—small ducks, shallow in breast and of narrow shape, bearing no resemblance to the grand Rouen or Aylesbury, from which in time gone by they possibly have sprung. It is the rarest thing to find pure a Rouen, Aylesbury, or Pekin on any farm, and it is a great mistake; for the young of these well known breeds will fatten up for table to 5 lbs. or 6 lbs. weight, without their feathers, in six or seven weeks with good management; whereas the farm duck can seldom be induced to lay on more than 3 lbs., and this at a larger cost of food, for the mongrel ducklings from a farm or elsewhere are of the greediest, and possess a most insatiable appetite. For egg-production too, the mongrel duck is useless. The eggs are small; and in November and December, when the early-hatched Aylesbury is hard at work, the complaint at the farm is that the ducks "dont't lay." The laying farm ducks are supposed to require no food, and if it were not for slugs and worms and grazing would pick up a scanty living.

FATTENING FOWLS.

This subject is described in great detail in many books and papers. The following hints may be useful to the novice in poultry-work. When a bird is to be put up for fattening it must be starved for some hours till very hungry. It must be fed during the whole fattening period very regularly at stated times; no food should be left after a meal; the coop must be very clean, and kept out of the sun, in a darkish place; the bird must not be allowed to see other birds running about, as this will make it uneasy, and retard the fattening process; the water must be kept pure and fresh.

No. 1. Fattening from the shell : Give the little chicks chopped hard-boiled eggs and bread-crumbs for the first fortnight after which add cooked meat chopped fine, and mixed with oatmeal and molasses. Corn steeped in milk is good later on.

No. 2. Take the birds when they have done growing; say at four months put them singly into fattening pens with barred floors through which the droppings can fall, and feed them with any of the following foods :—Oatmeal and chopped suet; boiled potatoes ; rice boiled in milk or molasses ; soft food of oatmeal mixed with milk or pot-liquor, pig's-fat, malt and sugar; suet is added the last week, and the birds are kept in sheds, airy but not cold.

ORDINARY FATTENING OF CHICKS.

For the amateur, the penning up and cramming system is not an absolute necessity. When the birds are from three to four months old, isolate them from all other stock in healthy quarters, with a moderate, but wholesome, run. Feed them every three hours with all the above foods in turn, and any other kinds of food you can concoct likely to tempt the appetite. Let them eat as much as they will, and be careful to remove what is left. Let the tins be kept sweet and clean. Give ample green food, fresh air and sunshine, with means of seeking shelter from its rays as well. Birds so treated will make healthy, firm flesh, and fatten quickly, but will not be laden with the soft fat which is laid on in the unhealthy air of the fattening pen. (*See* CRAMMING.)

FATTENING A LA BRESSE

The birds are kept on dry gravel, and fattened on buckwheat and maize meals ground fine as flour, and made into light porridge with milk.

Chicks should be ready without special feeding at four months old to kill, at from 3 lbs. to 4 lbs. weight.

FATTENING GEESE.

No amount of good food will fatten birds originally of a mongrel type. The quickest and best to lay on sound flesh are the produce of Embden geese crossed by a Toulouse gander. Do not proceed to fatten too suddenly. After giving free range of stubble and grass fields, confine gradually and at last wholly in a partially dark place. Wheat and barley grain and barley-meal with brewers' grains fatten well. Goslings may be put up to fatten at five or six weeks; seven weeks should bring them to perfection, Maize is also good, and turnip tops are greatly relished. Ponds are not required, but large troughs of water should stand about in the shade.

FEATHERS

are very valuable and where over a hundred head of poultry are kept, should be looked after as a source of income.

FEATHER-EATING.

A horrid practice, one might almost call it a disease, to which fowls brought up in confinement are liable. Malays and Houdans seem peculiarly apt to contract this habit, which dirt and crowding encourages. Idleness is one cause; poultry are often kept in a pen where they have no means of scratching about or amusing themselves. The earth should be forked up, thrown into heaps, and straw scattered over it; this will give occupation and tend to arrest the evil. Want of fresh water is another source of the disease; the water should be replenished often, and kept in the shade. Cabbages tied up whole and tightly to the walls of pens will amuse and serve to pass the time, and a piece of meat hung just within reach will be useful. Should any birds be so injured as to have the stumps of feathers bleeding these must be pulled out by the roots, and the tender places anointed with a salve of vaseline mixed with carbolic acid, 10grs. to the ounce. This will be healing, and at the same time unpalatable to the offending birds. Lettuce in large quantities should be given. If the case is desperate give daily ⅛ to ¼ gr. of acetate of morphia. The offending bird should be removed from the run.

LOSS OF FEATHERS.

Probably proceeds from deficient or unclean dusting arrangements. Fowls must have dust baths and one pound of black sulphur now and then, mixed with the rubble or sand, is excellent for keeping feathers in good order. A few grains of carbonate of potass in water twice daily and the application of petroleum ointment will produce a cure. Proper food is necessary for the preservation of plumage. Food without husk, such as Indian corn, soaked bread, if given exclusively, will bring on loss of feather; barley, buckwheat and barley-meal. If the skin is bare and shows no growth of feather, rub in oil and turpentine in proportion of three to one till the feathers break through.

FEATHERING OF THE LEGS AND FEET.

Perfection in this is no mystery, as amateurs imagine; it is merely a matter of extreme care. A bird to be shown with perfect foot-feathering must not once get into damp or mud, must not walk on stubble or rough earth, if not kept on a lawn it must have soft straw to walk on in the pen.

FELT.

Tarred is useful for covering roofs and buildings made of wood; it effectually keeps out wet and if tarred and sanded over now and then lasts a considerable time.

FENCING WIRE.

When putting up houses the lowest piece of wire should be of smaller mesh than the upper, so as to confine small chicks if necessary, and to exclude rats. Care should be taken *not* to nail the upper part of wire fencing to strips of wood; though neater to the eye it is better avoided, as the wooden strip, however light, gives the birds a foot rest; and they will attempt to fly on to it, which they will not do if it is merely strained wire at top. Asiatics will be kept in by a two foot wire, other breeds require it from five to seven feet high.

FLEAS.

A not infrequent but most unnecessary and disgraceful pest of the poultry yard. Whitewash three to four times a year, keep the places daily well cleaned, and such a visitation will seldom occur.

To destroy them, in an empty pen, paste up all cracks and holes, and burn a pound or two of broken sulphur in an iron vessel supported over water; let the fumes remain to do their work in the hen house all day, water the floor with carbolic acid solution; a wineglassful to the gallon, or pour turpentine in bad places.

Fleas get about the nests and building, rather than in the birds.

FLESH.

A decided aid to egg production, and during the winter months especially necessary, as there is no insect life. Scraps from the butcher without bone can be procured, these parboiled and chopped fine make an excellent dish. An exclusively meat diet on horseflesh, and the tainted carcases of animals unfit for human food, cannot improve the flavor of the eggs, and must be prejudicial to the health of birds. The French use horseflesh and a meat diet very largely.

FLOORS.

Boards are not good, especially if chickens are to be brought up upon them as they produce cramp. Concrete or brick is better, but nothing is so good as deep, well pulverized, dry soil, which is also less expensive.

FOOD AFTER EXHIBITION.

When the show birds return from their journey, place them in their own pen and give soft food only, warm—Spratt or bread and milk with very little water, as the birds; (being possibly feverish from long confinement) are apt to drink without measure and to a dangerous extent. Should the crop be overcharged with whole Indian corn or fresh wheat, a teaspoonful of gin will perhaps save an illness.

FOOD DURING MOULT.

Same as for ordinary adult laying stock, only a little more generous, and given warm; meat, ale, milk, and a little pepper, (red chillies are cheapest and best) may be added.

FOUNTAINS.

One of the cheapest and best home-made fountains is an ordinary flower-pot, reversed in a glazed saucer; fit a cork into a hole of the pot, fill it with water and reverse it. For adult fowls, tins which will hook on to the wire are excellent, as the water keeps clean in them.

FRESH BLOOD.

If birds are bred in and in to any great extent many evils will ensue—loss of size, fewer eggs will be laid, and a general want of stamina will be observable. It is well therefore, occasionally to purchase a cock from one of the best yards, and if it is for prize purposes, ascertain the pedigree, and, if possible, see the pen from which he was hatched. It is the easiest thing in the

LANGSHANS.

world to introduce a glaring defect into your flock, and one of the most difficult to breed a fault out. Where birds are kept in separate runs and pens, the produce for the following year or two will not be so nearly related as to require invigorating by fresh blood; in fact any large breeder of a well-known strain will be very shy of introducing new stock for any purpose. By a wise system of crossing and separation, thoroughly unrelated birds should be kept ready to hand for the mating season.

FRIGHT.

Sudden fright or much hunting about is the frequent cause of soft eggs. With prize birds it must be avoided, as beating about may cause irreparable mischief.

FROST-BITE.

Spanish, Andalusians, Minorcas, Leghorns and all large-combed breeds are specially subject to this. To prevent it, rub oil or vaseline over the comb with a sponge; but any fairly kept birds should not be subject to this danger. Rub, if frozen, with snow or cold water, and apply zinc ointment or vaseline.

GAPES.

Caused by pale, reddish worms, lodged in the windpipe, from two to twelve in number, and about half an inch long; each worm has a parasite worm attached to it. They kill the chick by at last crowding the windpipe till breathing is impossible. Dirt and damp is, as usual, the cause of this, as of most other diseases. If cleanliness and carbolic acid disinfection is freely carried out, gapes will be unknown. If cases occur, at once put fluid carbolate, camphor or lime, in the water. If there are many cases, place the chicks in a cold pit (garden frame) and fumigate with vapor of carbolic acid till they are nearly suffocated by its fumes. Care must be taken to liberate the chicks at the right moment or death will ensue, but if well done this is an effectual cure.

No. 2. Introduce an oiled feather into the windpipe, turn it twice and draw it out, when the worms will sometimes come with it.

No. 3. Give 1 gr. calomel, or 2 to 3 grs. of Plummer's pill, make the food hot with sulphur and ginger and wash the mouth out with chloride of lime.

GIDDINESS.

Produced usually by over-feeding; reduce diet and give aperient.

GRASS.

If free range in grass fields is an impossibility, arrangements must be made with some country children to bring in supplies, as it is very necessary for health and egg-production, not to mention that on its abundance depends the metallic lustre or gloss and firmness of feather which bespeak robust health and successful poultry management.

GRASS RUN

It is a good plan to combine hay-making with poultry-keeping. For six weeks before cutting, the birds must be removed to another field, or put up in other quarters. This gives time for the run to rest and be purified, and the birds so kept will be in more robust health than if kept from year's end to year's end on the same ground where only a limited space of grass is available, and several pens open out upon it. Each pen of birds must be let out in succession for a few hours' run, but at some period of the year the grass must have perfect rest. Grass runs are a luxury—*not* an essential—but the less grass the more care and work will be required to produce prize stock.

GRAVEL

A lime gravel is the best, and it should be screened to remove any large pebbles which prevent the birds from dusting with comfort. Every load of gravel should be mixed with fine sand or with fine mortar screenings.

GREEN FOOD

The best is grass, cut fresh daily; 2nd, lettuce; 3rd, spinach; 4th, cabbage, chopped in a chopper, or tied up to the wall or to a post for the birds to peck at. Every kind of green food refuse from the kitchen garden is invaluable, and a daily supply *must be found*. If on a free run of turf poultry will be seen to graze, like a flock of sheep. Hay refuse—out of a hay-loft, full of seeds—gives much amusement in winter time and is a good substitute when green stuff is low. Chickens should have grass, lettuce, and hay seeds in preference to other green stuff.

CLEANLINESS OF GROUND

Fowls will never touch food, if they can help it, which lies near any droppings or on an unclean place. Special care should be taken, therefore, to cleanse the ground of all pens and runs daily. The smaller the runs the greater must be the cleanliness add labor spent upon it.

GUINEA FOWLS

Are usually grey, having white spots on the grey ground; they are also pure white (more rare). If hatched on the land they will remain and roost with the poultry or on neighboring trees; but adult birds should never be bought, as they will inevitably fly away. The hens lay 100 eggs per annum, are very spiteful to the other poultry, and noisy early in the morning. They sit about August, and therefore it is wise to raise the chicks by artificial incubation, or by setting the eggs under a hen. Good chicken diet will suit the young ones but more meat is necessary. The hens like concealed nests, which should be provided, or they will lay away. They are quarrelsome, and it is only when an extensive range is available, that keeping the Galæna is found to answer.

HANDLING FOWLS.

If you catch a bird leaving its wings free a desperate struggle will ensue, likely to injure the exhibition plumage or to distract the broody hen from her vocation. Approach the bird from behind, place both hands firmly and quickly right over the wing-joints, then slip the right hand down and secure the legs firmly. All fluttering will thus be avoided, and the bird, held by the legs, with the left hand under the breast, will not offer resistance. All catching and handling of birds should be done at night, or after first making the pen dark, if this is feasible.

HARD FEATHER.

means the *close, firm,* appearance of a well-kept Game bird. To produce, feed on more grain than meal, and give plenty of run and green food.

HARDINESS.

Hamburghs and Dorkings are supposed to be delicate, and perhaps are rather more so than others; but with due care and shelter, most breeds will do well. Prevent overcrowding, damp lodgings and the necessity of standing about in mud and rain without shelter, and no breed need be considered too delicate. Leghorns and Andalusions seem very robust, and fear nothing, except frost so severe as to nip their large combs.

HARM TO CROPS.

Mechi, who speaks with some authority on farming, considers that poultry do no harm to growing crops on farm-land, that, at any rate, the damage to roots and seeds is trifling in proportion to the good they effect in destroying insect life, and the wire worm especially.

HAY.

Must not be used for poultry, as it generates vermin more readily than straw.

HEAT.

More dangerous in roosting places than cold; if artificial heat is given, it must be with thorough ventilation, or consumption will ensue, roup and loss of stamina. Heating of artificial mothers is much overdone, and too little ventilation given. The heat should not rise above 75 degrees when all are asleep, or suffocation will cause losses; for the first three days, however, chicks must be kept very warm, and will repay particular care in this respect. If birds are kept in confined pens, facing south, some shelter from the sun must be rigged up, tiffany or mats hung up for the midday hours; and the pen should be syringed with water; the fowls dust at once in the cool, sprinkled earth.

HEMP.

A great treat to all poultry, but if used too freely causes loss of feathers. A useful addition if the bird is out of condition, and where feeding up is required preparatory to showing.

HEREDITARY DISEASES OF FOWLS.

Consumption is the disease most carefully to be guarded against. A consumptive strain will be a constant source of care and disappointment. Squirrel tail is sure to be reproduced in many of the young birds. Wry tail is also hereditary. Crooked breasts. Thumb marks on combs, and any peculiarity in the spikes of the comb. White face, where red is the proper color is dangerously hereditary. Ear-lobes splashed or marked with red where pure white is a point. Vulture hock. All these defects will be reproduced. Birds with malformations, or anything missing, such as being short of one toe, or having any peculiarities should not be used for breeding.

HOARSENESS.

Birds occasionally, during a wet or hard winter, become hoarse the throat is evidently rough. Warm weather will remove this.

Glycerine and nitric acid in the drinking water will be beneficial.

HOSPITAL.

Every poultry-yard, in which, say, even 100 birds are reared yearly, should be provided with a place specially devoted for penning sick birds, where an invalid can be at once isolated and properly doctored if need be. This place must be open to the sun, screened from the east wind, dust dry, freely ventilated, yet free from draught, and warm. The hospital should be whitewashed with hot lime frequently, and perfect cleanliness maintained.

IN-BREEDING

Means mating the birds of one pen together, and these again with the cockerels and pullets produced by their eggs. This must be avoided, or the fertility of the eggs will be unsatisfactory, the produce will diminish in size, and the health of the strain will suffer. To avoid it, a cockerel or two should be bought yearly from other yards, or some pens must be kept so thoroughly apart that relationships will not be too close.

INDIAN CORN.

A favorite grain with all poultry but not good food unless the birds are on a wide range, being too fattening.

INDIGESTION

Shows itself by the birds going about moping, and disliking plain food. Give 5 or 6 grs. of rhubarb, and once or twice a grain of calomel; feed on cooked soft food, and let the bird walk about free, in a garden if possible, where it will pick up what is wholesome for itself.

INFLAMMATION OF THE LUNGS.

A sudden chill is the usual cause of this dangerous malady, and it is especially necessary to guard against it with artificially reared chicks very early in the season. Going out early in cold March winds after having nestled very warm (it may be too warm) under a hot water mother is most productive of this well-nigh incurable illness. A healthy strain will not be subject to this disease unless unduly exposed.

INSECT FOOD.

Wood-lice, worms, ants' nests, and maggots are excellent food for young chickens. Ants' nests can be procured by country children; and for a supply of maggots place some mealworms in an open shed in a box with part of an old mealbag; put in a dead bird now and then, and a regular supply can be kept up.

BREWERS' GRAINS.

Thrown in a heap and a dead bird or rat buried in them will also produce a supply of maggots.

JALAP.

One tablespoonful to twenty-five birds, in the food, promotes laying.

JOURNEYS

If long, are best accomplished by night. The darkness conduces to sleep and quiet, and an extra day of confinement and loss of food is not incurred. Twenty-four hours before the bird starts it is well to write a line to advise the recipient of the fact as delay is thus prevented.

KILLING FOR TABLE.

This is done in various ways, but whatever be the manner let the birds fast for at least fourteen hours before death. One of the most humane methods is to tie up the legs and bind the cord twice or thrice round the wings, and then with a very sharp axe to chop off the head on a wooden block. Draw the skin at once over the stump after the bird has been hung up by the heels to bleed thoroughly. Another way is to cut the jugular vein and bleed well.

LACED FEATHERS

Have a narrow border round the edge different in color or shade from the ground color.

LAMENESS.

Examine the feet and if no cause can be discovered then consider whether there are too few hens to each cock. This will produce a lame or feeble way of walking, which should at once be prevented by the addition of more hens to the pen.

TO SECURE EARLY LAYING.

Hatch early, and do not move pullets about to various runs while they are maturing. Do not depend on old hens, but on March pullets, kept in warm quarters fed on meat and plenty of green food, besides grain; occasional treats of bread soaked in ale, hot, and a few pounded chillies mixed in the food, and given hot at daybreak will hasten the filling of the winter egg basket.

LAYING MIXTURES.

There are many mixtures and condiments advertised in the poultry journals daily, which have the effect of stimulating the hen's laying powers if desired, but few will be needed and many are prejudicial; and if the above diet is kept to, the birds must lay; if they do not, either suspect and look out for rats, or egg eating hens in the flock, or a need for padlocks to the laying pens. A very effective egg-producing mixture is one oz. of Glauber salts given in a meal of potatoes three or four days running, 1 oz. to ten fowls; but this must only be tried on common laying stock. It is effective, but dangerous, and must be used rarely and with caution.

TO PREVENT LAYING.

Birds for show have at times to be kept back. They are in show form just when they begin to lay, and never look so well after. If you are early and wish to delay the laying and to prolong the period of growth move the pullets about from one run to another.

LEG-WEAKNESS.

Chickens brought up on boarded floors, without a free run or exercise, are subject to this; or if kept in heated places or on too stimulating diet. To prevent it from a week old give free range with beds at night of dry, sandy earth to sleep upon. Feed with bone-dust mixed in meal, and give chemical food in the drinking water or thirty drops twice daily, of the syrup of hypophosphite of soda.

LIGHT-COLORED YOLKS.

These show an insufficiency of iron and sulphur in the food, also absence of green food.

LIME.

An absolute necessity in the formation of the egg-shell, and for the proper action of the digestive organs. No fowls should be left without it. Any builder will supply old mortar and rubble for the cost of carting. This must be screened and placed in dry covered runs.

LIMIT TO NUMBERS.

It is pretty well ascertained that fowls do not succeed if kept in too large flocks, however extensive the range may be. If kept in separate flocks of fifty or less they pay and answer better.

LIST OF FOODS.

GRAINS.—Barley, buckwheat, hemp, linseed, millet, maize, oats, Scotch groats, wheat.

MEALS.—Barley-meal, Indian corn meal, linseed, oatmeal.

AIDS IN FEEDING.

Bone-dust, Spratt's Food, rice, oyster shell, salt, meat, house-scraps, bones, liver, fat, fish, hay seeds, sun-flower seeds, red chillies, etc., green food of all sorts.

If feeding is carried out with judgment, and changes are rung upon the foods above named, the poultry will be kept in health, and will produce chicks with good, vigorous constitutions.

LIVER DISEASE.

Indigestion is often disregarded until it develops into disease of the liver. The birds mope about, show an irregular appetite and a bilious yellow hue appears in place of the coral-red which a thoroughly healthy bird should show in face, comb and wattles. Doses of aperient medicine, preceded by a grain or two grains of calomel, will at times effect a cure if the case is not of long standing.

LOBES, OR DEAF EARS.

Are the folds of skin, either pure white or red, cream color or blue, according to breed, which hang from the true ear, large in some races as the Spanish, small in others, as in Dorkings. Whiteness is best preserved by keeping the birds out of strong sunshine. Fighting may produce an unsightly scar, of the lobe if cocks get together. Oxide of zinc in powder dusted over the lobe will preserve it from the effects of hard weather, and zinc ointment, (benzoated) will be found useful in softening and preserving the texture in show birds. Should red ticks or spots appear, give less exposure. If brown spots appear on lobes originally pure white, they may be from too rich feeding. Reduce the diet, and doctor the lobes with glycerine and carbolic acid (5 grains to the ounce), or with sulphurous acid in water, 2½ drachms to the ounce.

LOSS OF APPETITE.

When this occurs suddenly, give half a teaspoonful of Epsom salts; this will often start laying.

MANURE.

Poultry manure should always be saved; it is most valuable if kept dry. A water-tight tub with a cover should stand at the door of each block of poultry-houses, into which the droppings can be thrown as free from sand and extraneous matter as possible. A little earth or ashes sprinkled over it will effectually deodorise, and do no harm.

MARCH, WORK FOR.

Hatching must now be in full swing, as it is important to get the bulk of all valuable stock out before the 31st. Stock for table may be hatched throughout the year. Allow chicks all the freedom possible; do not shut up in small places, or cramp will result; the more they run the better. Eschew board floors. Give plenty of milk. Chicks are much helped by warm food. It is of more importance to give food warm, sweet, fresh, and very frequently, than to pamper the appetite with condiments and luxuries. Stock pens should be let out in the grass runs whenever the weather is not bad. Stimulants should be dropped by degrees if they have been resorted to; but the clever poultry-keeper's account for these will be small. Preserve eggs from frost if for hatching. Put false eggs into the nests to encourage laying and sitting. Birds must be put under restraint in wet and severe cold; our climate will not allow of free range in winter. The birds will *not* really range, but will stand about and get chilled and wet; hence much of the failure in poultry-farms. Supply large covered runs, open to air and sun-gleams, but dry.

MARKING CHICKENS.

It is possible with the hot point of a needle to perforate the wing membrane with one, two, three, or more holes, marking the various hatches or strains. Colored cotton round the leg answers, but must be changed as the size of the leg increases, or it will lame the chick by growing into the flesh.

MATING.

If early chickens are wanted, mating should be arranged by the end of November. Pullets and hens over their moult will begin laying about a month after mating. Any great faults in the hen must be counteracted by the influence of the cock, or the fault will be exaggerated in the progeny. All birds with marked defects must be excluded, and only the best bred from. When once mated the pens should not be disturbed, as any change may be the cause of unfertile eggs.

MAY, WORK FOR.

Hatch on still for late shows, and for supplying successional pullets. Do not spare the sulphur dredger, and water the nests with hot water frequently if the weather is dry. Elder broods should all be sorted, cockerels to one run, pullets to another; and more air should be given at night, while rain and wet grass by day are not so likely to hurt the chicks as the weather grows warmer. Plumage of last year's show birds will now grow brittle; guard birds intended for show from storms, mud, and strong sunshine. Dust heaps and baths should be cleansed, thoroughly renewed, and well mixed with sulphur powder. Unlimited green food should, as usual, be given to old and young stock. The food for Turkey chicks should be frequent and plentiful. Potatoes, barley, wheat, oatmeal, turnips, Spratt, milk, rice boiled; soft food mixed with plenty of lime and brick-dust will fatten and keep them in health. They are very susceptible to damp during the first week, and should be protected from showers.

MEALS.

Number in the day for adult birds three. At daybreak in winter, at 6 a. m. in summer; at midday; and at 5 p. m.

MEAT PRODUCERS.

For delicacy only, Game is the first. For delicacy and size, Dorking, La Fleche, Crevecœur, Houdan, Langshan, Malay, and Brahma-Dorking with their crosses.

MILK FOR CHICKENS.

Boiled milk given warm is decidedly good for chickens, and prevents diarrhœa. It is indispensable for prize-bred chickens, and should be given at least every night and morning. Nothing is worse than allowing milk to stand about and get sour. All milk pans should be scrubbed twice daily with hot water.

TO HASTEN MOULT.

Pen up cocks apart from hens in a warm place, with deep sand and mortar siftings. Keep them very warm at night; the older the bird the warmer it should be kept. The process of moulting takes about two months, but at times much less. Food should be given warm, very little at a time, and not stimulating when first penned up; then generous diet, with a touch of red chillies in it; and in the gallon of drinking water put sulphate of iron the size of two filberts, and ten drops of sulphuric acid. Non-sitting hens can be hurried on by taking away all stimulating food and placing them in a fresh house. As soon as moult begins feed well. Should birds moult too slowly, and look ailing, give two or three one-grain doses of calomel, a dose of jalap, soft food, and meat. If the weather is cold, a pill now and again of two or three grains of cayenne is useful. The Spanish tribes moult late and hard; birds with shabby feathers in July cannot be ready in time for September exhibition. If required early they must be preserved from injury, for moult cannot be hurried on so early in the season. Meat, green food, a little pepper, and ale, with warm housing at night, will bring all birds comfortably through the moulting season. If the shaft of the new feathers seems to stick on for a long time, not splitting open freely, more stimulating diet should be allowed; meat, linseed, and hemp.

NEST EGGS.

Should be of china or painted wood. It is a very slovenly and bad plan to put addled eggs about as nest eggs; if broken they will pollute the nest and hen's feathers also, and render her otherwise than sweet and wholesome for the next valuable eggs she may be required to incubate.

NESTS FOR LAYING.

Hens like to be secluded when they lay and sit; nests affording shelter from the vulgar gaze will therefore be preferred. The nest should not be so formed that the hen must jump into it; this shakes the eggs at times so violently as to break them. A box with the bottom and one side taken out will form a snug nest, if put next a wall and kept well supplied with fresh straw chopped in short lengths. If rats abound raise the nests from the floor. The wire basket nests sold in various sizes suited to the different breeds are excellent. These are hung on the walls of the house, but if put high a board must be fixed to the wall hard by the basket, so that the hen can get quietly into the nest without jumping in. Change the straw or whatever be the material often—hens appreciate cleanliness— but do not change the position of the nest, as it gives offence to laying hens, and they may punish you by laying no eggs for a day or two.

NIGHT ACCOMMODATION FOR CHICKS

Must be dry, warm, ventilated, and secure from rats, very clean, and supplied with broad, low perches or shelves, not touching a wall, covered with straw.

NUMBER OF HENS TO COCK.

Five to one cock if the eggs are being sold for prize stock at prize stock price; but if for ordinary farm purposes, eight, or even ten on a free range.

NUMBER OF DUCKS TO DRAKE.

If there are twelve ducks three drakes would be advisable, and so on in proportion; but the eggs are more fertile if a drake and four or five ducks are kept in separate flocks, for in a large flock the drakes are apt to be quarrelsome and interfering.

OATMEAL.

Scotch is the best; though expensive it pays for prize poultry; the birds get on wonderfully when fed upon it as a staple article of diet, mixed dry with rice boiled in milk till it crumbles into a fragile mass.

OATS.

Too much husk to be a favorite grain with poultry, but very good in change with other corn.

OLD CHICKS.

The first brood of the season is sure to get especial and individual attention in the way of constant feeding, delicate diet, replenished hot-water mothers, and various delicate attentions. Care should be taken as the numbers increase that the interest does not flag, and that younger chicks are not left on short commons in the way of food, tit-bits, warmth, and all the care without which they will not be a success.

OLD FOWLS FOR TABLE.

Nothing is more trying than a tough old hen for dinner. The poultry papers give invaluable receipts for rendering old members of the poultry yard as tender as the young ones. For this very long and gentle stewing is necessary. An old hen stuffed with mustard, salt, and pepper, is excellent.

For an old stager of four years the following is advised. Kill her, pick, and wrap while warm in vine leaves, then bury her, and let her lie for twenty-four hours in sweet earth. Lastly, boil very gently in good stock, and the result is tenderness as of a chicken.

ONIONS.

Almost all green food is valuable for poultry, and should be collected carefully and thrown to the birds fresh daily, but onion-tops and onions must be carefully excluded, as hens are fond of them, and the eggs will taste very strong and unpleasant after they have eaten them.

ORCHARD.

Orchards are far preferable to open fields for poultry farm purposes; the shelter of leaves in summer is very beneficial. Cankerworms and caterpillars falling from the trees are consumed, windfalls are made use of instead of harboring vermin, which again creep up and destroy good fruit.

OVER-FATTENING.

Care must be taken not to keep the birds penned up for cramming a day after the process is complete, for after a certain time disease and emaciation will set in.

PACKING EGGS FOR HATCHING.

Sudden or sharp jerks and jars are to be warded off by the packing medium. For thirteen eggs, get a box, inside measure about 12 by 6 inches. Sew a strong piece of canvas to the edges of the bottom of the box, leaving the canvas loose, so that it can be filled up with hay pad. Wrap each egg in paper, then pass round it a wisp of hay. Proceed to bed the eggs in, small end uppermost. with hay between, and as you pack fill up all spaces with chaff; lastly, put a layer of hay at the top, and screw the lid on the box. The address should be clearly written, with—"Eggs for hatching." "Not to be shaken." "Immediate."

PAINTING FEATHERS, LEGS, ETC.

An elaborate species of deception practised by some dishonest exhibitors. Leghorns' legs have been found to be painted or dyed of a bright yellow color.

PARALYSIS

May be complete, or only of the legs. In either case it is impossible to cure thoroughly, and the bird will always have a lame or awkward gait. Unless a peculiarly valuable prize bird, it will be best to kill it. Daily faradization might be useful, and might be carried out on a valuable bird, with the aid of gentle doses of strychnia when the case is partially recovering.

PARASITES.

Parasites should not exist, and their presence in any number shows great want of cleanliness. If whitewash, dust baths, sulphur, and Persian powder, etc., fail, apply petroleum ointment under the wings, about head and inner part of thighs, but this is a very severe measure. Syringe the house with hot water, in which carbolic acid is dissolved, a wine-glassful to the gallon.

A tobacco leaf in the nest will drive off insects, and keep laying nests free of these pests. A few insects may collect under the wings and under the thighs of birds, especially those which are a little out of condition and chickens reared by broody hens, which are collected haphazard from neighboring farms, but they ought quickly to be got rid of by the clever poultry keeper, and should be the exception, not the rule. There is nothing better than Persian powder mixed with sulphur, used out of a dredger, for powdering chickens; a dash of it over the head and neck,

GOLDEN SEABRIGHT BANTAMS.

another between the legs and under each wing, will clean the chick in a day. It is very necessary to pen the chicks in a bare pen devoid of straw for an hour or two after the operation, and then when let out boiling water should be poured over the ground where the insects have fallen; this prevents their recovery and cleanses the place.

PARSNIPS,

Boiled, are good for poultry-food, and assist laying.

PEA COMB.

Such as is seen in both varieties of Brahama; the comb is in three ridges, the centre one rising above the outside ones, all distinct, and all firm, and compact, rising from the front and arching back.

PEKIN DUCKS.

The largest ducks known. Chief characteristics; brilliant orange bills and feet; white plumage, with the under parts of canary hue; boat-shaped bodies, and a strikingly erect, penguin-like carriage. They do not seem to reach such heavy weights as the Aylesbury or Rouen, though larger to look at. The first prize pair at Birmingham, 1879, scaled under 14 lbs., while Aylesbury and Rouen reached a weight of over 22 lbs. in 1878. They fatten quickly, and are a contented and quiet breed. It is possible to keep them in confinement, and to get plenty of eggs with a fair percentage of ducklings, even though the parents have no more water than is supplied by a rain tank sunk in the ground four feet by six. This can be done, but free range with ponds or a brook is, of course, better still for breeding stock. Pekins are not fanciful as to food, take willingly all that is given them, and the ducklings fatten quickly to a large size.

PEN FOR BREEDING STOCK.

Should have a grass run attached besides the covered run, where the birds may be left in peace to go in and out as they like. The less they are fussed the more fertile will be the eggs. For sale of prize eggs at high prices put four hens to the cock; for farm work, or for house supply of eggs and chickens, eight to one cock. If only for egg production, not hatching, any number to one cock.

PENCILLED FEATHERS.

Have no moon, no border, but dark bars in parallel lines across lighter ground.

PERCH.

Allow six inches for each fowl. The best is a pole sawn in half, with the knots cut off smooth, the bark left on. The perch should not be placed high when there is no space for the birds to fly down with a gradual swoop, as is natural to them when roosting in the open; three feet from the ground in confined places is high enough, the perch should be far enough from the back wall to keep the tail plumage clear; carelessness in this causes much mischief. Broad, low perches should be erected in sunny spots about the run; fowls delight in perching to preen themselves after meals.

PIP.

A name given to a dry scaly substance on the tongue of sick birds; it comes when the breathing is obstructed through the nostril, and if the bird's health is restored the pip will vanish. The mouth may be cleaned with Condy or chlorinated soda.

PLUCKING

Should be done while the bird is warm. An excellent plan in small establishments, where there is small accommodation for feathers, is to have newspaper bags made and when a fowl is plucked let the feathers be put into it. These can be at once baked in the oven, which will destroy any live stock, and the store of feathers may be put by, in a compact, cleanly way, for sale when the season is over; a collection of dirty feathers in any other way is very objectionable. Pekin and Aylesbury Ducks' feathers are very valuable, as also is goose down, which fetches the highest price.

PLUMAGE.

The thing to be aimed at is a close, firm, plumage, with a brilliant gloss upon it; grass runs will give this. To attain it in confinement exceeding care, extreme cleanliness, and clever management in feeding, are necessary. Hemp and sunflower seeds are excellent for imparting a glossy appearance; too much must not be given, or feathers will drop off.

PRECOCIOUS CHICKENS.

None are more so than the Andalusian, Minorca, and Leghorn. Cockerels of these breeds will crow at two months, and call the other chickens to feed, giving up to them delicate morsels like gallant old birds. It is well in such cases to separate the cockerels from the pullets.

PREVENT HEN SITTING.

Put the hen into a new run of poultry; the change of cock and companions may have good results, and the loss of the favorite nest at the same time completes the cure. Dipping hens in water and putting them into solitary confinement on short commons is cruel and unnecessary.

PRIZE BIRDS, THEIR TREATMENT.

Hatched in the three first months of the year, they must be well fed and well housed, never chilled, and yet allowed perfect freedom on the grass runs whenever fine and dry. Soft food should have bone-dust mixed with it, and the meals should be ample and frequent, but never so large as to remain uneaten and to get sour; meat and green food should be given in plenty; at from three to four months the cockerels should be separated from the pullets; no crowding, no want of cleanliness should be allowed, and no roughing it in bad weather, or the feathers will be soiled. These must be kept spotlessly fresh, and care must be taken that no rough wire, or ill-made doors or awkward perches, injure the plumage, on which prizes to a great extent depend. Three weeks before the show, pen the birds, cock and pullet separately, giving each a friendly companion of their own sex; feed on bread-and-milk, wheat, and twice or three times a week give linseed; boil it till it is in a jelly and mix in oatmeal till it is friable; this will gloss the plumage. Also give barley-meal and sharps, buckwheat, a little hemp, oatmeal-and-milk, with a little meat. Let the pens be deep in fresh straw; see that the dust-baths are very clean. Two days before the show give night and morning a meal of rice boiled in milk, and plenty of wheat; a little meat chopped into the rice is much enjoyed. Rice is to prevent any chance of diarrhœa in the show pens, which would entail extra soiling of the plumage. Green food is to be given in plenty, preferable grass and lettuce and spinach. Forty-eight hours before showing wash the birds if need be. Feed as above till an hour before starting. Lastly, wash the comb, face, etc., with soap and water, dry it, and rub it over with a little vinegar; give each bird a teaspoonful of port wine; they will then sleep instead of fretting in the railroad car. Inside the basket at the side tie the top of a loaf soaked with port wine, and a lettuce, to peck at; this will bring them in good spirits and condition to the show pen. Do not omit, three days before the show, to give the cockerel or cock a hen in his pen, but not one which is to be exhibited. He will then not take much notice when the show pullet is introduced into the exhibition basket, and this should be done about three hours before the train leaves to insure that no fighting occurs.

PRIZE POULTRY.

Does it pay? There is no doubt that keeping poultry for exhibition and the sale of thorough-bred stock is remunerative if you can combine it with the sale of eggs and table poultry. Prizes bring no grist to the mill, the expenses of exhibiting being great, while the enormous charges of the railway companies consume the profits of even first-class prizes.

PRODUCE HATCHED, GOOD AVERAGE,

Out of first-class prize eggs: chicks seven out of twelve traveled prize stock eggs: twelve first-class birds fit to show out of two hundred chicks, Spanish; eight first-class birds fit to show out of two hundred chicks of other breeds. Another great breeder says he has nineteen walks of Dorkings, cock and four hens in each; by the 1st of April he has 172 chicks, by 1st of May 400. If he gets fifty prize birds fit to show, and many of these to win, and 150 fair birds besides, he considers it a very lucky year. If properly managed, an egg farm could be very well kept up at the same time that breeding for exhibition is carried out. Exhibiting is expensive work, but it is necessary, that the prize stock may become known; when this is accomplished it may be discontinued as an unnecessary expense. To pick out six or ten birds fit to show and win entails the hatching of say two hundred birds annually. Those falling short of the show standard can be drafted into the laying or fattening pens at once.

PULLETS.

The pullet is so called for twelve months, or until the year in which she was hatched is closed. Pullets hatched in April, 1880, for instance, would go through all the shows of that season, from July to the following February or March, as a pullet, and so with the cockerels. Pullets should not be mated till they are five months old, and then with an adult cock rather than with a cockerel. That their eggs do not hatch is an error; they do so as well as those of older birds, but the produce is not quite so vigorous unless the pullet was hatched quite early, in February or March.

PULLETS NOT LAYING.

If over six months old they are either over-fed, which can be ascertained by feeling their condition and weighing; or possibly underfed; if pullets are much exhibited and the runs often changed this will prevent egg production. Should the birds be thin give meat and a little stimulant, as red chillies or condiment, buckwheat, sunflower seeds; if fat, reduce diet and give an aperient. Constant exhibiting is very fatal to laying.

PURITY TO BE PRESERVED.

Unless the accommodation is very ample, it is a great mistake, when going in for prize poultry, to try several breeds at one time. Amateurs are fond of doing this. The danger of mixing the breeds is too great, and should not be lightly run when eggs are sold at prize prices.

QUANTITY OF FOOD.

It is impossible to feed fowls or chickens by measure. Never leave food to be trodden about, let all be eaten up clean.

REPLETION.

Some birds will over-eat themselves, and mope about after meals in a dejected manner. This may be the fore-runner of more severe disease from over-feeding.

RHEUMATISM

Shows itself in the same way as in human beings; stiffness of joints, contraction of the toes, and a painful gait. Warm quarters; a hot bath for the legs, which should be bandaged with flannel, and rubbing the legs with chloroform or soap liniment, are useful; give half a grain of opium five times in the week; and good generous diet. No one would take this trouble, save in the case of a very valuable exhibition bird.

RICE.

An excellent food for chickens. It must never be given raw, but boiled well until soft, and in skimmed milk if possible; if not, dripping or fat should be added, or coarse sugar for a change, and then dry oatmeal should be mixed in till it is a crumbling mass. It may be given in turn with oatmeal and Spratt's Food.

ROOFS.

The best material is corrugated iron, for then the rain-water from the roof can be collected and stored. A wooden roof covered with tarred felt is good also, but not so lasting, and the water which runs off must not be given to the poultry to drink, neither can it be used for watering flowers, as it is highly injurious. Slate or tile roofs are equally good. Thatch should be avoided as it harbors rats and mice, and unless very thick will not keep the wet out, especially if deep snow melts on it.

ROOSTING.

Chickens should not be allowed to roost till from four to five months old, and then on broad perches, two to three feet *only* from the ground. Exhibition birds are better roosted on shelves covered with sand and littered with straw, but then they must be kept very clean, and the straw must be frequently changed. On no account place perches one above the other so that the droppings fall on birds lodged lower down. Cleanliness is the great key to success, and the roosting-places should be scraped out daily and re-sanded. The greater the number of fowls, the greater must be the care.

ROUEN DUCKS.

In plumage exactly like the Mallard or Wild Duck. For exhibition the drakes must have the breasts rich red-brown, of darkish hue, the drake's bill yellow with a greenish tinge, not lead nor bright yellow, the bill to come straight down from the head, long, broad; the legs, rich orange, and the head rich glossy green, and round the throat is a ring of pure white, but this must not go right round; the back is greenish black; tail, darker; wings, grey and brown, and a bar across of brilliant blue, edged with black-and-white, clean cut; the flights are grey and brown; the fluff and under parts must be toned down to light grey, *no* white must be seen. The duck's bill, orange color, must be nearly covered, but not to the tip, with an irregular splash of dark color, blackish; the ground color dark, chocolate brown, with pencilling of still darker tint.

Birds for breeding are good weight at seven pounds. In the show pen they have exceeded twenty-four pounds, and were once shown over thirty-two pounds, but such fattening destroys breeding power, and the birds are useless. The eggs are not so large as the Aylesbury; they are both colors, green and white, and very plentiful; the flesh is as good as the Aylesbury, and they fatten equally well.

ROUP.

This dreadful disease beging with a slight cold, followed by inflammatory symptoms. It affects the cavity of the nose. The discharge commences by being watery and clear, afterwards becoming thick and offensive; the face, eyes, and throat swell; lastly, fever, thirst, and loss of appetite, come on.

No. 1. At once isolate the bird when as yet nothing but a slight cold has appeared. Wash the head with warm milk; feed with soft food. Powder the roosting places with quicklime just before they go to bed, and give daily one grain sulphate of copper in oatmeal; give warm ale and plenty of green food.

No. 2. Foment the swelled parts, squeeze out the matter; give teaspoonful castor-oil, and feed on oatmeal with pepper; plenty of green food.

No, 3. Drop solution of atropine into the eyes when running. Give a pinch of Epsom salts daily, soft food, and a capsule of cod-liver oil with quinine. Squeeze the inflamed glands, and dress with one part Wright's Liquor Carbonis to fifteen parts water.

No. 4. Keep warm and dry; give meat-scraps daily; apply alum and cayenne, a little in mouth and nostrils, as snuff, to cause sneezing and clear away mucus; bathe the nostrils with Condy or carbonate of soda, and give daily a pill of meal with two grains myrrh, five grains carbonate iron, two grains cayenne.

Should any bird get a severe roupy cold, it is advisable, unless a very valuable bird, to kill it at once, and so stop the spread of an infectious and troublesome illness, which *may* clear out your stock.

RUNS.

For prefection poultry should have grass, earth, and gravel on their runs. The covered run should be on the original soil, but well and deeply covered two spades deep in a mixture of gravel, sand, old mortar, and road drift, all screened. In this the fowls will delight to dust, and vermin will not irritate them or their owner. Grass range is of the utmost value, and should be let out as soon as the dew is a little cleared off. Earth and manure heaps are invaluable for scratching in, and to supply animal food in the way of worms, etc. But the birds will not thrive, however extensive the grass range, if they are not provided with a light, open run, roofed in, and free from drip and damp, wherein they can freely dust themselves and keep in shelter during rain. Birds will not often shelter, however wet the weather, if they have to do so in a dark, damp, and dirty, airless house. If they do, it is to mope about in idle discomfort, which brings on evil habits and illness of all kinds.

Yard Run. If a bricked or paved yard is the only place available in which to give the poultry a run, great results must not be expected, but eggs and healthy birds may be secured with attention to the necessaries of poultry health. A load or two of screened rubble, gravel, etc., must be at hand under shelter for the fowls to dust and bathe in. Access to a manure and garden refuse heap is highly advantageous, giving occupation in way of scratching and hunting for the animal food which it engenders, such as worms and insects of all kinds. Green food must be thrown to the birds daily, and the manure must be as regularly cleaned of. In the absence of these precautions, fowls can scarcely be kept in health in a paved yard.

Earth Runs. If the space is only moderate, earth runs are superior to any. A small grass run soon gets used up and becomes foul. Beyond sweeping it nothing can be done, and the grass soon loses its freshness, whereas an earth run can be raked and swept daily, and dug over three times a week. Twice a year it should be *double-dug*—that is, the soil should be taken out, two feet deep, in a trench at one end, and carried to the opposite end. The trench should then be filled up as the digging proceeds. Thus fresh soil is brought to the top, and thorough cleansing is secured.

Covered Runs. An absolute necessity for perfect health and exhibition condition. Each should have a large trap-door to the free range, and a well-made roof, impervious to rain and drip, and the dusting material should here lie deep, fine, and clean.

SAND.

Useful to mix with all dusting materials; it should be kept dust-dry to mix with earth and to form bedding for the artificial mothers.

SELF HELP IN BUILDING.

When building poultry-houses, on which most people grudge

any large expenditure, one can often dispense with the aid of a carpenter or builder. Carefully plan and measure your intended buildings, and get the wood sawn to proper lengths at some dealer's or sawmills—any handy man with some little help should then be able, if interested in his work, to put the houses up. Care should be taken to let the boards overlap, and to char or tar all posts let into the ground.

SOUR MILK.

Sour milk is an excellent drink for young chicks. Let them have it to run to. It is also good for laying hens. The great trouble is that many poultry raisers can not get enough of it.

SAWDUST.

A bad packing medium for eggs, the jar of the railway journey shakes it away from the eggs be they packed ever so firmly. Good for pigeon lofts if used deep and in quantity, but not for poultry.

SELECTING EXHIBITION BIRDS.

A very large number of birds must be hatched from which to make a selection. Any with glaring disqualifications must be drafted out early in the season. A little later another lot must be cleared out of those with faults, but not sufficient to disqualify, leaving, say, twenty birds out of two hundred hatched. To these give every possible advantage in the way of space and food. Some will answer expectations, and some will fail. Amateurs are often too sanguine, and imagine all are going to be prize-winners, whereas it takes no little care and experience to attain to the much-coveted honor. Birds have a different look when penned up and when free in the fields. Choice should be made when the birds are in their pens or runs, and great care must be used to match the pullets with the cocks, a well-matched pair being considered most desirable in the show-pen. When choice is once made, the birds should not be sent about to small shows, but should be reserved for one of the more important exhibitions.

SEPARATING COCKERELS AND PULLETS.

This must be done at three, four, or five months, according to the breed, some being far more precocious than others.

SEX OF EGGS.

That this can be foretold is an old woman's tale ; it is certain, however, that the first batches of eggs in the early season chiefly produce cockerels, and that five or six pullets mated with adults cocks produce pullets in greater numbers than cockerels, while from a vigorous cockerel mated with about three or four adult hens, cocks will be in the greatest proportion. There is no way of discovering the sex of an egg before hatching.

SHEDS

Are invaluable as shelter from rain and mid-day sun. If such erections exist already on a farm, the roosting poultry houses should be built on to them ; the expense of covered runs can thus be avoided.

SHELF UNDER PERCH.

When birds are kept for laying purposes, this is highly advisable. The collection of droppings on the boards placed to catch them reduces the trouble of daily cleaning. as with an iron scraper it is the work of five minutes to scrape off the droppings of fifty or a hundred fowls. It is a good plan also for prize birds, as far as cleanliness goes, to have a board under the perch to catch the droppings, but this shelf must be carefully adjusted as to its height from the perch and from the ground; it should be at such a distance from the perch that the birds may walk under the latter without injury to the sickle feathers; some hens squat on the board, instead of taking to the perch, and even in one night this may prove fatal to show plumage.

SHELTER HURDLES,

Thatched with straw, and placed on four posts eighteen inches from the ground, afford admirable shelter from the extreme glare of the mid-day sun, and also places which they can retire to in winter from the damp ground, and preen themselves in the short hours of sunshine.

SITTING HENS, THEIR MANAGEMENT.

During January, February, and March, it is most difficult to get broody or sitting hens, yet this is the best and only time to hatch the bulk of exhibition stock, as well as the pullets which are to fill the egg-baskets in August, September, October, and November, when eggs are highest in price. For good sitters seek out in preference Brahmas (especially the Light), Cochins, Dorkings crossed with Brahma, and Silkies. Moderate-sized hens have an advantage in not being so heavy. Hens are to be preferred to pullets. Eggs are most fertile in March, April, and May. In January and February they are the most valuable, it being necessary to hatch for early shows. Let the hen sit, if possible, where she has chosen her nest (this should be a movable one, either box or basket), and give a few china eggs to experimentalize upon. While she is off feeding, clean out the nest, place in it moist sifted earth to a depth of three inches, and on this make the nest of chopped straw (about six inches long). Let the hen return to her nest of her own free will; then in an hour or so, when she is firmly settled, gently carry her, covered up in the nest, to the sitting pen, coop, or wherever you mean her to sit, and if in a few hours' time she is still quiet, give her the eggs she is to incubate; after this do not disturb her, even for feeding, during the next thirty-six hours at least. The nest should be placed in a pen or coop by itself, where no hens or chickens can enter; give food (barley), and water, and a barrow-load of dusting material, and leave the hen alone to come off and feed when she likes. This is the best method; but when some thirty or forty hens are to be set it is difficult, in most cases impossible, to provide separate pens, and it will be found that if many hens are set in one place, to come off at their discretion for food, great confusion and fighting will be the result. Two hens will get into one nest, leaving others empty, to the fatal injury of the eggs. Where a sitting-house is employed it is well to have the nests in rows round the walls, in which the hens should be shut up, each in her box, and taken out together every twenty-four or thirty-six hours to feed, after which they should be replaced and shut in. The best nest for this purpose is a box with the lid on hinges, and one side taken out; this open side should be placed over the earth and straw nest, the lid should be perforated with holes for air, as also the upper *side* of box to give free ventilation; in this way a large number of sitters can be housed in a small space. If the sitting-house has an outer pen or run where the hens are fed, so much the better, as it will then not disturb the sitters so much. Great regularity in the time of feeding is requisite, and extreme gentleness and quiet. Fresh water, plenty of barley and wheat, green food, and dusting dry material are necessary. If the latter be mixed with sulphur to keep the sitters free from vermin, so much the better; it will ensure *quiet* sitting, for a hen tormented, with vermin will be restless. If you suspect she is so worried, put about half a pound of powdered sulphur all over the nest before you put the eggs in; no vermin will stand it, and the effect is marvellous. If hens have to be procured from strange yards, the removal and sitting should be carried out at night, and the hen should keep her nest twenty-four hours before the eggs are given; with these precautions it is quite possible to get good sitters from a long distance, even five and six hours' journey by railway. The number of eggs to be put under a hen varies according to the size of the hen and of the eggs, also according to the season; during January and February not more than eight or nine should go under a large hen, and after that from thirteen to fifteen. While the hens are feeding it is well to pour hot water round the nests after the first week, to cause a moist heat when the hen returns to her labor of love. After the fifth day the eggs can be examined with an egg tester, and unfertile eggs replaced, these must be extra new laid, so as to hatch out within two days of the others; three or four days before hatching is due eggs may be floated in a bucket of water at 105° Fahr. All eggs which do not bob about or rock to and fro with a pulsating motion may be discarded. Eggs will sometimes remain in the water three or four minutes before the movement is noticed, or they may move or pipe at once; they often chip in the water, and must instantly be lifted out. In this way more room is given to the hatching chicks, and the risks of a broken addled egg or crushed chick in the nest from over-crowding are avoided. If the weather is frosty a handful of hay put over eggs when the hen feeds is useful to keep in the heat, and the hen may remain off safely for fifteen minutes. In very dry weather, besides pouring water round the nest, a little may be sprinkled on the eggs, but the hen must be at once replaced, or mischief will arise from a chill. If a hen forsakes her nest, and eggs are found cold, place them at once in water at 105° Fahr., and leave them in till you provide, as quickly as may be, another nest and a hen willing to become a foster-mother. Eggs neglected and chilled should never be despaired of; it is a fact that eggs left over twelve hours, and stone cold to the touch, when treated as advised, have hatched eleven out of thirteen, and all the chicks strong and healthy. As the chicks hatch out it is well to place them in a drying box till all are out, by which all risk of the hens squeezing them will be avoided. The same person should always attend to the "sitters," and extreme regularity, gentleness and regularity in management, has much to do with success in hatching.

SLIPPED WING.

This chiefly occurs with fast-growing cockerels and ducklings. The primary feathers, which are naturally tucked up out of sight, stick or trail out; the bird has no power to tuck them up. Should the same feathers stick out and appear twisted, so that the inside of the quill is outside, it is probably an hereditary evil. In the first instance it frequently occurs from a number of cocks being kept together, giving rise to some ill treatment, constant racing about, and nervous flapping of the wings; these being soft and delicate as yet, the birds fail to fold them in closely, and a habit is acquired of letting them hang down out of place. Tucking them up into place when the bird is asleep at night is sometimes effectual; but the best way is to sew a band round the wing-feathers near the shoulder, and attach this to another which is passed round the joint of the wing to prevent it slipping off. It is a work of patience and difficulty.

SNOW.

Poultry will eat quantities of snow. They should be shut up if seen pecking it, and melted snow-water is not good for their drink. When shut up at these times, a few handfuls of hayseeds, the sweeping of the hay-lofts, afford excellent amusement for them.

SPRING CHICKENS.

Hatch them in November, December, and January, and have them ready for table February, March, and April.

SQUIRREL TAIL.

When the tail rises up very upright from the back, and inclines to bend towards it instead of taking a graceful sweep from it.

STAINS ON FEATHERS.

To remove these use turpentine or benzoline, but the best plan is care and cleanliness to prevent accidents. Cleaned-up birds never equal well-kept birds on a grass run.

STRAIN, TO COMMENCE.

The best plan is to purchase a pen of the breed you desire—say a cock and two hens—from some well-known breeder, paying a good price, and trusting to his sending birds properly mated for the purpose you require, whether for showing, or for raising thorough-bred stock from which to select future exhibition birds. Having done this about October or January, procure in March a sitting of the very best eggs of the same breed from the next best breeder, unless the first can be depended on to send you some not related to the birds already purchased from his runs. Hatch these, or as many sittings of these as you conveniently can, and select the best of the chicks for stock, mating in the following December your bought cock with the best hatched pullets, and your best hatched cockerel with your bought hens and two pullets to make up the pen. Another plan would be to go to some of the principal shows and buy birds there if you are a good judge of the breed. This may answer, but there is a risk, as you cannot know the antecedents of the prize-birds, and may be introducing fatal faults into your strain to start with.

STRAW.

Never mind the untidy appearance, and let it lie about in dry covered runs.

SULKY COCKS.

Towards autumn cocks sometimes become very cross to their hens, and a trial to the whole harem. Isolate them at once, they require rest and quiet; feed well and house warmly till over moult. In December mate the pen up for the season, and the most irritable bird will have become good temper itself.

SUNFLOWER

Seeds are useful to give gloss to the feathers.

TONICS.

Quinine and iron tonic (citrate of quinine and iron): four grains to an adult fowl daily.

Sulphuric acid ten drops, and sulphate of iron a piece the size of a filbert, in a quart of water for drinking.

Tincture of iron, one teaspoonful in the quart of water.

Nitric acid acts on the liver, and is a tonic. Of the dilute acid four drops, in a teaspoonful of water three times a day, or ten drops of strong acid in a quart of water for drinking.

TESTING EGGS.

Eggs should be tested either on the sixth or seventh day, this applies to both those used in incubators and under hens. The method of testing is to grasp the egg with the thumb and forefinger of the left hand, hold the egg up between the light and the eye, shading the light from the eye with the right hand. If the egg is cloudy looking it is fertile, if perfectly clear unfertile. At seven days a fertile egg will have a black spot in the centre, gradually shading off to the edges where it is much clearer, whilst an unfertile egg remains clear all the time.

The easiest method of testing, is to buy an egg tester, they are sold at a very low price and to those who have not tried testing before are more certain.

TRAVELING.

Birds travel best by night. If a long journey, feed at dusk —meat and corn—give water, and start them off as soon after as train will suit, giving a teaspoonful of port wine to each bird on starting, and tie into basket bread and lettuce.

TREATMENT AFTER EXHIBITION.

On the arrival of birds from an exhibition, feed them on soft and (if cold weather), warm food, with a very little water, containing a tonic. See that they are housed *very* warm. If they are shortly due at another show, give bread-and-milk for one meal daily, and rice-and-milk, with meat, and, if possible, a grass run. If the crop is loaded with Indian corn, feed very sparingly even on soft food at first; and if it feels *hard*, give a teaspoonful of gin on arrival; it will aid digestion.

TRUSSING.

There is great art in trussing, and amateurs often fail to sell at profitable prices owing to the slovenly way in which their birds are prepared for market. When killed, fowls should be plucked at once, and placed on one of the boards for shaping dead poultry for table used in France. It may improve the appearance of a lean, ill-fed chicken to break the breast bone and hide it away, but there it still is, destroying the breast slices and a constant trial to those who have to carve the bird at table. If properly fed and plucked. and pressed on the trussing board, this operation is quite uunecessary.

UNFERTILE EGG.

Clear eggs, which, after three or four days' incubation show no appearance of fertility, no veins, or sign of embryo. The yolk will be seen to float and ocillate with every movement of the hand. Such eggs must be turned out of the nest or incubator at the sixth day, and are perfectly good for the most deiicate cooking purposes.

VENTILATION.

A neglected but most important subject. Poultry-houses are, as a rule, either draughty, or they are unventilated; if the first, the birds are always uncomfortable, and a late egg supply, owing to cold housing, will be the result; if the latter, serious disease will follow, such as diptheria, or the birds will be dull, without appetite, the wings will droop, upright combs will go blue at the tips, and fall over limp and flabby. Besides the door entrance, every roosting-house should have a window, which can be left open on hot nights, a wire of small mesh should 'bo placed over it to keep out enemies; in the winter a piece of perforated zinc is preferable, as it prevents the wind rushing in, and yet gives enough air. If a window is not practicable, a hole under the eaves will answer, covered with zinc wire; the higher up ventilating openings are made the better. Foul air rises, and openings must be made, or the fowls will suffer. Ventilating holes should be drilled in all artificial mothers, dryers, and shelters; foul air generates very quickly where chickens congregate.

VERTIGO.

An ailment which betokens over-feeding to an enormous extent, and a threatening of apoplexy; a bird so afflicted will stagger about in a circle as if giddy. Dash cold water on the head at once, and give castor-oil, jalap fifteen grains, and two grains of calomel. Feed very little, reduce the system; the bird is sure to be a very fat one.

WANT OF CONDITION.

A bird shown badly—*i. e.*, dirty, wanting in gloss, with ruffled plumage—would be passed over, unless his points were quite super-excellent, and over-showing will destroy the good looks of the finest specimens.

WARMTH.

Comfortable housing is all the warmth required by adult birds. Artificial heat is dangerous, as it is difficult to control. Heated houses are a fertile source of consumption and many other evils. A cock and a few hens might, however, be forced for the supply of early eggs, which certainly come the sooner for extra warmth. In artificial mothers the heat should not be kept above 70° when all the chicks are collected under them asleep.

WASHING EXHIBITION BIRDS.

Get two tubs, fill the smaller one with a good lather of soap

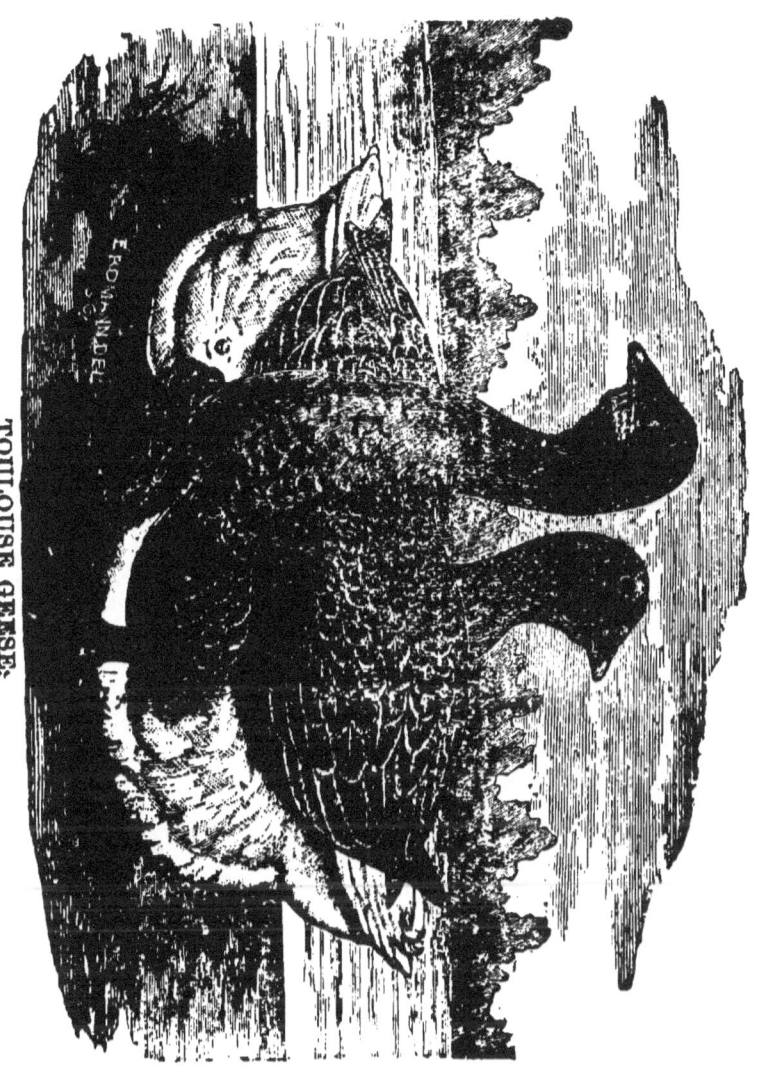

TOULOUSE GEESE.

water (for one bird half a pound of white soap is sufficient); stand the bird in the lather and wash it, using a softish hairbrush, and with it your hand; thoroughly brush and cleanse the feathers everywhere, leaving no spot untouched, and do not be afraid of wetting thoroughly, use no half measures; take care not to bend or brush the feathers the wrong way. This done, having prepared warm water in the larger and deeper tub, dip the bird in and out freely and thoroughly, rinse every vestige of soap lather out; lastly, take a can of merely chilled water (may be *very slightly* tinted with blue for white birds), and pour this over the bird, drain and dry as far as you can in a Turkish towel, place the bird in an exhibition basket, from which the lining has been half removed, and set it at a comfortable distance from the fire. The basket should have half or three parts of the lining left round it to keep off draughts. As the birds dries and fluffs out, gradually draw away from the fire; leave the birds the night in a warm kitchen, and next morning place them in their own preparing pen, which, meantime, has been laid deep in fresh straw; let them rest here for twenty-four hours, or twelve, at any rate, before the journey, otherwise a risk of cold is incurred. After the bath, when still wet, give a teaspoonful of port wine, and later a meal of bread and meat scraps, which are gratefully devoured as a rule; by-and-by a handful of groats as a treat cast in the straw will tempt them to scratch for it. A moist, warm atmosphere must be kept up in the drying basket, or the feathers will not web properly; place water within reach, and add to it a little tonic.

If the birds are not drying properly, try and turn them about so that the heat will strike all sides equally. Hard-feathered birds, such as Andalusians, Minorcas, Brown Leghorns, Malays, Dominiques, Game, Black Spanish, do not require so much washing. White birds and Asiatics demand the greater care.

WASTERS.

By this is meant those birds which, although pure bred, and hatched from prize stock for exhibition purposes, fail in some important point. They should at once be banished to the fattening-pen, or be killed for chicken-pies.

WEAKNESS AFTER SITTING.

When it exists shows neglect and want of food; the hen lies about, pecks at grit and green stuff, but rejects food. Slip a raw egg down the throat, give half a teaspoonful of quinine wine and oatmeal food warm. If she will not pick it up, make a bolus of it, dip it in milk, and cram her daily.

WHEEZING AND COUGH.

When the birds are not otherwise ill, do not be anxious, but put a teaspoonful of glycerine and a few drops of nitric acid in the water.

WIRE SUPPORTS TO COMB.

When a cock's comb is given to lopping, or hangs over, unprincipled exhibitors have been found to stick needles or wires through the flesh to keep it up. Such wicked cruelty would of course disqualify. A wire support can be made to be worn outside the comb, and this is quite allowable with a view to making it grow straight; it may be of some help, though a really good breeding or exhibition bird should be quite independent of such artificial aids.

WORM EATEN FEATHERS.

Hamburghs are subject to this disgusting disease, if badly kept, but it comes only from unpardonable neglect, damp, dirt, and deficient arrangements for dusting. Treat as for "Lice," saturate the quills with oil, cleanse the birds and their home in every possible manner, if not too far gone for recovery.

CAPONIZING.

The custom of caponizing is practiced quite extensively in Europe, but is poorly understood in America. We see no reason why it can not be made very profitable to those who will only take the trouble to learn the *modus operandi* and practice it.

When it is made plain that a cockerel at the age of four months, by caponizing, can be made to increase in size and weight, at ten or twelve months, nearly twice what he would be, it is a surprise that more market poulterers do not practice this art, in a country as progressive in most things as this one is. But it seems that the people generally, in America, treat with indifference anything pertaining to poultry, save in a small number of fanciers, scattered here and there, who are striving to beat it into the heads of farmers and stock-growers, that there is something else besides laboring in the hardest manner from sun to sun to earn a dollar, or that there is other stock of importance than cattle and hogs.

The mode of proceeding to be successful, is to take the young cockerels when they are three or four months old, keep them from feed the morning you intend performing the operation so that the bowls are empty, for, if they are fed, and full, you will almost be sure to cut the intestines while making the incision and cause death

"The instrument employed in the operation should be very sharp; a surgeon's small operating-knife, termed a curved-pointed bistoury, is far better than an ordinary knife, as it makes a much neater wound, and so increases greatly the chances of healing; or a curved-pointed penknife may be used. A stout needle and waxed thread are also requisite; a small curved surgical needle will be found much more convenient in use than a common straight one.

"It is necessary that there should be two persons to perform the operation. The assistant places the bird on its right side on the knees of the person who is about to operate, and who is seated in a chair of such a height as to make his thighs horizontal. The back of the bird is turned towards the operator, and the right leg and thigh held firmly along the body, the left being drawn back towards the tail, thus exposing the left flank, where the incision has to be made. After removing the feathers the skin is raised up, just behind the last rib, with the point of the needle, so as to avoid wounding the intestines, and an incision along the edge of the last rib is made into the cavity of the body sufficiently large to admit of the introduction of the finger. If any portion of the bowels escape from the wound it must be carefully returned. The forefinger is then introduced into the cavity, and directed behind the intestines towards the back, somewhat to the left side of the middle line of the body.

"If the proper position is gained (which is somewhat difficult to an inexperienced operator, especially if the cock is of full size), the finger comes into contact with the left testicle, which in a young bird of four months is rather larger than a full-sized horse-bean. It is movable, and apt to slip under the finger, although adhering to the spine; when felt it is to be gently pulled away from its attachments with the finger and removed through the wound—an operation which requires considerable practice and facility to perform properly, as the testicle sometimes slips from under the finger before it is got out, and, gliding amongst the intestines, cannot be found again readily; it may, however, remain in the body of the animal without much inconvenience, although it is better removed, as its presence is apt to excite inflammation.

"After removing the left testicle, the finger is again introduced, and the right one sought for and removed in a similar manner. It is readily discovered, as its situation is alongside of the former, a little to the right side of the body. Afterwards the lips of the wound are brought together and kept in contact with two or three stitches with the waxed thread. No attempt should be made to sew up the wound with a continuous seam, but each stitch should be perfectly separate, and tied distinctly from the others.

"In making the stitches great care should be taken; the skin should be raised up so as to avoid wounding the intestines with the needle, or including even the slightest portion of them in the thread—an accident that would almost inevitable be followed by the death of the animal.

"After the operation the bird had better be placed under a coop in a quiet situation, and supplied with drink and soft food, such as sopped bread. After a few hours it is best to give him his liberty, if he can be turned out in some quiet place removed from the poultry-yard, as, if attacked by the other cocks, the healing of the wound would be endangered.

"After the operation the bird should not be permitted to roost on a perch, as the exertion of leaping up would unquestionably injure the wound; it should, therefore, at night be turned into a room where it is obliged to rest on the floor previously covered with some clean straw. For three or four days after the operation the bird should be fed on soft food; after that time it may be set at liberty, for a short period, until it has recovered entirely from the operation, when it should be put up to fatten."

ARTIFICIAL INCUBATORS.

The great question of the day, with people about going into the poultry business is about Incubators. "Are they a success?" "Which is the best one?" "How do you work them, etc. ?" You will find the poultry papers full of these "questions," and the "answers" are sometimes quite amusing. One distinguished "poultry fancier" has never tried them; another, (not quite so distinguished) has failed to hatch more than twenty-five per cent. of the perfect eggs put in, and still another, a fortunate man truly, hatched ninety-eight per cent. of eggs perfect and imperfect. Truly the days of sitting hens are numbered.

But for all these facts in favor of or against the Incubator, it is a settled thing that artificial incubation, attended to by a proper person, will and does pay. Inventors have been taxing their brains for many years to produce good machines, and after many failures, this year of 1884, finds at least a half a dozen good ones in the field competing for popular favor.

The bad ones are legion, and as Halstead remarks in his work on "Artificial Incubation," that many of the so-called Incubators, are not much better than a tin pan with a lamp under one end and a tray of eggs under the other. Some of them being perfectly worthless for the purpose, and only a fraudulent means of obtaining money from a class of people who always make "cheapness their one criterion of value."

The best way *we* know of to select an Incubator, is to note their advertisements in the poultry papers, write to the manufacturers and obtain their circulars and testimonials. If they have taken premiums at any of the State fairs within the past year so much the better, they are worthier of consideration. Examine testimonials, and if you see the names of noted fanciers appended to them, men who are known throughout the land, by their successful yards or by their articles in the different poultry papers and magazines, you are safe in trusting the Incubator manufacturer. It is only the testimonials containing unknown names and out-of-the-way places that are to be distrusted. I shall give in the following pages, engravings of some of the noted Incubators with the modes of working them, I do not do this to advertise them, but merely to present to the eye of the "uninitiated," the "machine" in all its glory; the world is wide, there are many of them, you can take your choice.

(*For the following articles on Artificial Incubation we are indebted to Mr. J. L. Campbell, of West Elizabeth, Pa.*)

ADVICE TO BEGINNERS.

"I am often asked the question, do you think I can make a success of poultry raising with an Incubator and Brooder? Now that is certainly a difficult question to answer. How am I to judge of the abilities of one I know nothing about? Raising poultry is just like any other business, one will succeed where another will fail.

"I will answer that question by asking several more. Did you ever raise any chickens? Do you like to work with and care for them? Are you willing to give them the necessary amount of care? Do you expect the business to run itself, or do you intend to run it?

"If you can give satisfactory answers to all of these questions, I can tell what you can do.

"In general terms I will say to all who desire to go into it as a business, if you don't know anything at all about the business, and don't intend to devote the necessary time to it to learn it thoroughly, you had better keep out of it. To fanciers and all others who understand the raising of chickens, I would say, if you only want to raise a few hundreds, even for the early market, it will pay you well to get an incubator, the profits of a single hatch has in many instances paid for the machine.

"Those makers who tell you that no care is required, that the business will take care of itself, that all you have to do is just to spend ten or fifteen minutes each day attending to the incubator, etc., all they want is to sell you an incubator, after that you can whistle for the balance.

"The real trouble about artificial raising is that so many have an altogether wrong idea about the care of the chicks, and that all that is necessary is to hatch them out, and then throw them a little feed occasionally. Well, a greater mistake could not be well made; any one who goes into it with that idea can either expect to change the idea very suddenly, or quit the business. Now, the facts are just this: The real work comes in when the chicks are out of the shell; then is when the most care is required. It is no trouble for one who is used to the care of chicks, and knows just what they need, to succeed; but for a novice to go at it with the idea that they can do it is as well as any one, is all a mistake. To all such, I would say, either go in with the idea that you are going to give it the care and time to learn and a few failures will not discourage you, or else keep out of it altogether. There is money and plenty of it in the business, either for market poultry, or eggs, or both.

Poultry raising is a business that is especially suitable for women. There are thousands of farmers' wives and others, who, if they would try the artificial raising of poultry, could make it pay handsomely. These women all know more or less about the care of poultry, and could make it a success from the start. Women generally understand the raising of poultry better than most men; one reason is that they attend to the numerous small details which are necessary, but few men are willing to give the time that is required."

BEST PLACE TO RUN INCUBATORS.

This will depend somewhat on the kind of an Incubator it is, a good Incubator can be run with success in almost any kind of a place, but of course it would always be best to have the conditions as favorable as possible.

The very best place that could be selected is a room that would have a temperature the year round that would not go below sixty or above eighty degrees.

If the room could be kept so that it would not fall below sixty and not use fire, it would be the best to never have any fire, there is no trouble in keeping up the heat in the incubator (if it is a good one) even at a zero temperature, but the trouble is that if you have it in too cold a place you cannot take out the trays without chilling the eggs.

If the eggs can be turned without taking them out of the machine, it will not make as much difference as it would with one that the eggs would have to be taken out frequently in order to turn them, but even then it would be necessary to take out the eggs to test them and for that reason the room should not be too cold.

Therefore the best place to select when it could be done, is a room on the north side of the house where the sun would not strike it too hard in hot weather, it is more difficult to make a good hatch in very hot weather than it is in cold. Then, if it is desired to hatch in cold weather, the room should be warmed up to at least sixty degrees when it is desired to take out the eggs either to turn or to test them.

A real dry, well ventilated cellar is a good place to hatch in, either in winter or summer, but it should be well ventilated and perfectly dry, else it will not give good results, the ventilation is next in importance to the heat, no matter where the incubator is, a good hatch will never be obtained if the air is not pure.

The very best temperature that a room could be kept at is seventy-five degrees, this gives enough difference between the temperature inside the machine and out to insure good ventilation, and there would never be any danger of chilling the eggs when they were taken out.

The hardest time to make a good hatch is when the temperature of the hatching room runs away up into the nineties, the worst trouble about it is that the eggs and the outside air is so near even in temperature that there is very little ventilation.

I have made hatches in hot weather when the machine would have to be left standing open all day, and never have the lamp lit at all during the last week, the eggs keeping up all the heat that was needed, I tried the eggs on the fifth day after the lamp was out just to see how high they could run up the heat in the incubator, and they actually raised it to one hundred and eight degrees, of course there was a large number of eggs, over one thousand, and the temperature of the room at the time was ninety-eight degrees. I have no doubt at all but what it would have went still higher, but I was not willing to risk it any longer for fear of killing all the chicks, of course it could not have went much higher as just as soon as there would be a few degrees difference between the inside and outside of the incubator the air would begin to circulate and keep down the heat, but it shows what a large amount of heat is in one thousand eggs when the chicks are nearly full grown.

This is one reason of many failures that have occurred with incubators, enough allowance was not made for the animal heat that was in the embryo chicks.

A good incubator or rather one that has a good regulator will be able to take care of itself during the first two weeks, even if the temperature should range as high as ninety-five degrees in the shade, but during the last week, if the weather should go above ninty degrees the incubator will have to be watched during the hot part of the day no matter how good it is, for it may turn off the heat and open the ventilators and still the heat may go too high. To sum up then, those who want to run an incubator and not give it only the least possible attention must get a room that will be cool in summer and warm in winter.

When the weather is at all cool there is no necessity for taking out the eggs to air them, in fact, it is much better to not do it, if they are cooled down to about ninety-five degrees two or three times during the hatch it will answer every purpose, I have hatched ninety-eight per cent. of the fertile eggs and not let them cool off once during the whole hatch.

I made a test of six trays, one hundred and seventeen eggs in each tray, the eggs were kept at one hundred and two to one hundred and four degrees all the time, that is the regulator was set to run from one hundred and two to one hundred and four degrees and back again, the change would occur about once an hour on the average, these eggs were out of the machine once only and that was on the sixth day for testing them, the result was that I had from one hundred and four to one hundred and thirteen chicks out of each tray, the weather was cool and pleasant, neither warm nor cold.

I have made a great many tests with the same end in view, *i. e.*, to find out whether it is necessary to air the eggs or not, and I find that in cold weather it is best to not air them at all, in cool, pleasant weather it is not necessary, but will do no harm unless they get too much of it, in very hot weather they should be aired daily.

If they are not aired in hot weather the chicks will be weakly, and if they are aired frequently, in cold weather the result will be just the same, weak chicks from too much chilling.

Eggs can be hatched at a temperature that will vary from ninety-nine to one hundred and ten degrees, that is they can be hatched in large numbers, I have tried some at one hundred and twenty degree for a short time and still hatched some of them, but all such chicks are worthless they will neither live, or, if a few of them does live they will not grow or do any good.

But if they are hatched at an even and proper temperature and all the other conditions to successful incubation are complied with, the chicks will be much stronger and more healthy, and will outgrow those that are hatched by the average hen.

Those that are hatched at too low a temperature will be weakly, if hatched at too high a temperature they will be weaker still, overheating makes the blood turn to water, if it is carried too high the red color will entirely disappear, of course the chick is dead before that can occur. but they can be heated hot enough to entirely destroy the vitality of the blood and not immediately kill the chick. Therefore, above all things else, avoid too much heat, the proper temperature will receive attention futher on.

AVERAGE TIME REQUIRED TO HATCH.

The average time for all breeds is twenty-one days, the actual time for the different breeds is from nineteen to twenty-two days. I have frequently hatched an occasional egg on the seventeenth day, but I am convinced that the eggs were either started before they were laid or else they had been sat on by a hen before they were put in the incubator.

If is a fact not generally known, that a hen will sometimes carry the egg after it is fully formed, until it will begin to hatch before it is laid, but this is a rare occurrence, and can never take place under ordinary circumstances, for the hen nearly always drops her egg in a few hours at most, after it becomes perfect, but it sometimes happens that a hen from fright or other causes will retain her eggs until she has three or four perfect eggs in the oviduct all at once. I killed a large hen once that had seven perfect eggs in her and she was unable to get rid of any of them, three of these showed that incubation had started quite plainly.

Sometimes a hen will be so badly frightened that she either cannot or will not lay the egg at the proper time, and in that case the egg production will cease, now what is to hinder the hen carrying the egg two or three days and then dropping it, and what is to hinder it hatching afterwards, and if it did it certainly would have that much of a start. It is a well-known fact that the germ cannot begin to develop until it enters the perfect egg, then what is to hinder it starting to develop in the hen as well as out, some will say, why, want of air of course, but that will not hold water, all those who know anything about the structure of the body of the hen knows that there is a large open cavity, and they should know that this cavity must be full of air, else the body of the fowl would collapse instanter.

Now, who knows what kind of air this is, whether it contains oxygen or not, I will have to confess that I do not, but would like to know, but I do not see why it should not, and if it does what is to hinder the egg having enough air in it before it is laid to start it hatching while it is still in the body of the hen? That it does do this sometimes I know.

Any one who will take the trouble, can see the air bubble in the large end of the egg just the instant it is laid, although it will be quite small at first, so that the air that is in it when laid must certainly be there while it is still in the hen.

I should, therefore, conclude that all eggs that hatch under nineteen days, have been started one way or the other, before they were put in the incubator.

The average time of Leghorns, Games and Bantams in a good incubator is twenty days. The average time of Cochins, Brahmas and all the large, slow moving birds should be put at twenty-one and one-half days; I have hatched large numbers of Cochin eggs and the great majority of them came out on the twenty-second day; Plymouth Rocks hatch in just twenty-one days; Houdans, Polands, Spanish, Hamburgs, and the Dorkings, all hatch in twenty-one days. Ducks and Turkeys in twenty-seven and one-half days, and Geese in twenty-eight days

The fresh eggs will hatch quicker than stale ones, and also will hatch more healthy chicks; I hatched a Plymouth Rock egg once that was laid on January 2d, and I put it in the incubator on the 18th of March; it had a small hole in it where a little nail had penetrated the shell, I stuck a bit of paper over the hole; this egg had laid in a drawer in the office of a store room all the time. The chick was perfect, but it did not live, it never eat any, and was weak when hatched.

Eggs that are well taken care of from good stock will hatch well until three weeks old, middling well at five weeks, but longer than that I would not care to hatch them, because, even if the chicks do hatch out they are not so healthy and vigorous.

It will depend greatly on how they are kept, my plan is to stand them with the large end down, and keep them at a temperature as near fifty-five or sixty degrees as possible; some advocate keeping the small end down, but I never could see the philosophy of that method, the main object in keeping eggs, is to prevent the yolk adhering to the shell, and I can see no better way to make it stick fast than to stand it on the small end, while, if it is stood on the large end, the air bubble will tend to keep it off the shell, while the pressure of the yolk will tend to keep the bubble from spreading and drying up the egg, while, if you stand it with the large end up, you place it in the best position for drying up, and, also for to stick the yolk fast to the shell, the very two evils that you are trying to avoid.

It is known to all that have studied the structure of the egg that the large end is the most porous, any one can test it for themselves, by holding a stale egg under water, and all the air will come out of or near to the large end, now the inside skin is not fast to the large end of the egg, it is loose nearly one-third of the way down, and if it is allowed to fall away from the shell it will just allow it to dry up that much faster, while, if it is held down to the shell it will keep the pores closed and prevent the drying up process.

In hatching chickens, those that come out on time will always be the best and strongest, those that come out later than the twenty-third day, are usually not of much account, if they have been kept back for want of sufficient heat, they may come out and do all right, but if it is because the egg was not good and the germ vigorous, it would not pay to try to raise them, overheating will cause them to be late in coming out, in that case the chicks are useless and will not pay for raising them, **even if they live at all, which will be doubtful,**

I have tried how long I could keep eggs back by keeping a low heat, and I have made them hatch on the twenty-sixth day, but I never yet could succeed in getting a chick to live that was later than the twenty-fourth day in coming out, and seldom then.

CARE OF ARTIFICIALLY HATCHED CHICKENS.

There is no subject connected with the business, that deserves the attention that this does, for on it all the balance depends.

The most important requirment of young chicks, is warmth, the next is an abundance of pure air, and to be kept perfectly dry, then comes the feeding, and unless this is properly done, all the balance will fail, the feeding will be treated of separately.

Chicks when first hatched should be kept at a temperature of ninety-eight to one hundred degrees for the first twenty-four hours, and great care must be used to not allow them to get overheated, it is much easier to overheat five hundred or one thousand chicks just out of the shell than it is to overheat that many eggs, and it is just as fatal to the health of the chicks to overheat them after they are hatched, as it is while they are in the egg, the only difference is, that it will take a little more heat to kill them than it will while still in the egg.

They require no feed for the first twenty-four hours, although it will do no harm to give them a little if it is desired to do so, for there is no danger that they will eat enough to hurt them.

Then, as soon as they are large, or rather old enough to be fed they must be divided in small lots, if a large number is fed in one lot there will always be some of them that will not get any, the outside limit that ever I could succeed in feeding properly in one lot is one hundred, fifty is better, and twenty-five is just that much better still, those who divide their chicks up in flocks of twenty-five will stand a much better chance to raise all of them than they would if left in larger lots.

Then it is very important to select out all the weak ones and put them by themselves, these, if left with the others will soon be trampled to death, while, if put by themselves will soon pick up and be all right. I refer to the small chicks and not to those that have any thing the matter with them, a chick that is sick when hatched out will never come to anything any way.

And just here it is pertinent to remark that healthy chicks can only be hatched from eggs that are laid by hens that are in vigorous health, if eggs are hatched from yards that have the cholera the chicks are pretty sure to have it before they are a month old.

There is no one thing connected with the raising of artificial chickens that require more care than to see that they don't pile up and smother, the best remedy for this is to keep them so warm and comfortable that they will not want to pile up, chicks that are sick or weakly will be much more apt to do this than those that are in vigorous health.

The only way to do it is to keep them in a brooder that will keep them just warm enough and not too warm, and it must give them an abundance of pure, warm air, even in warm weather chicks require a little artificial heat, at night, not because they realy need it, but because if they do not have it they will be sure to crowd and smother.

It is best that the heat should be above the chicks, but while it is warm above it should not be cold below for if it is the chicks will not be comfortable.

The best plan that I have ever tried is to have a tank of warm water over the chicks lined with flannel underneath and in addition to this a heater to deliver pure, warm air directly in over the backs of the chicks, they are so comfortable that they do not think of crowding.

No matter what plan is used there must be abundant ventilation and no draughts of cold air.

They must have sunlight, the more the better, but it is pretty safe to say that it would be impossible to raise a flock of chicks to the age of three months and not let them see the sun, I cannot do it, whether others can or not I do not know.

WATERING.

Another important thing is the watering. Some claim that no water at all should be given, and others claim that they should have it all the time. But I have always had the best success by not doing either the one or the other but by adopting the middle course, my plan is to give no water in cold weather except what I put in the feed, and in warm weather I give them from one to three drinks daily owing to how warm it is, if the chicks are at liberty they can have water all the time and it will not hurt them, but when confined as they must be in cold weather, if it is before them all the time they will drink too much and it soon weakens their digestive apparatus so that they soon depart for another world.

They cannot live without water, but it can be given in the feed in such a manner that it will furnish all the moisture that is required by the chick and at the same time not hurt them.

FEEDING.

On the proper feeding of chicks the success or failure of everything else depends, one great point is gained by starting right.

There is no better article for the first feed than good, sweet, stale bread and milk, made just wet enough so that it will be soft.

Now, I wish all who read this to bear in mind that I am

giving directions for the feeding of chicks that are in confinement, if they are at liberty and have a good, large range they will not require near as much care as they do when confined and will do well on a much more simple diet.

Well, as I said before, begin with stale bread and milk, this can be fed several times at first, all the eggs that are just the least bit addled and all the unfertile ones that do not hatch (an unfertile egg never hatches) can be used, eggs should never under any circumstance be fed dry, they should be boiled until quite hard and then mashed fine, shells and all, and should then be mixed with something else, either bread, potatoes or oatmeal, and then made soft with sweet milk, but never sloppy. All feed must be cooked, a certain amount of cornmeal is good, but it must never be fed raw, there are many ways to prepare it so as to make a good, wholesome food. One good way is to scald the meal then to mix it in a stiff batter, use sour milk, and add enough bi-carbonate of soda to the milk to make it foam, a teaspoon to the quart is the right quantity, then salt slightly and add a small quantity of lard, mix well and then bake in a hot oven, but not so hot as to burn it.

When the weather is cool, enough can be baked at one time to last several days. the best way to feed it is to crumble it and then slightly dampen with sweet milk.

The best wheat bread for chicks is made from whole wheat ground fine and then sifted to take out the coarse bran, bake it just the same as if you were going to eat it yourself.

Then, for variety, use potatoes, rice, barley, and a little fresh meat of some kind, meat should be cooked and then cut fine, very little at a time must be fed, they should have a little once a day, two or three bites for each chick is enough.

The best meat chopper that I have ever tried is that made by the Enterprise Manufacturing Company, of Philadelphia, Pennsylvania. It can be obtained of nearly all hardware dealers, it is cheap and easy to keep clean and just what a raiser of poultry needs. The man who invented it deserves a gold medal a yard in diameter.

The best way to feed potatoes is to boil with the skins on and then mash up fine, for small chicks the skins should be removed, then add a little salt, plenty of black pepper, a teaspoon of pepper is not too much for one hundred chicks, and if the potatoes are very dry add a little sweet milk, they can also be mixed with other articles of food, the best way to feed the meat is to mix it with something else.

The potatoes must never be watery, nothing but good, sound and sweet food of any kind must ever be used, nothing will destroy a lot of young chicks quicker than sour feed.

For green food there is nothing superior to onions and cabbage, give me these and I will agree to keep the chick in good health and not use any others. The onions must be cut fine and they can either be mixed with the other food or they can be fed alone.

In feeding, variety should be the rule, give a different food each time if possible, never give more than they will eat up clean, never leave any food to become sour and dirty or you will soon spoil the appetites of the chicks so that they will not eat and then you are done, a chick that will not eat is not of much account.

The main object in feeding is to get the chicks to eat all the feed that it is possible for them to eat and remain in good health and condition, and the best way to do this is to not leave it before them all the time but to give it often, my rule is to feed three times daily, some successful poultry raisers feed five times, giving a little at a time and it is a very good plan, but I think that my plan is the best, I give three main feeds daily and then I allow them to have something to peck at between times to amuse themselves, a head of cabbage, some apples, sweet apples are much the best, ripe tomatoes are good when they can be had.

Bones and oyster shells are two of the best articles for chicks that can be used, but it depends somewhat on how the bones are treated, they should be obtained while fresh and sweet, and then place them in an oven or a stove and roast them until they are quite brown, it is best not to burn them black, they will then pound or grind up quite easily, and when roasted will keep sweet for a month, keep some of this and some pulverized oyster shells before them constantly, the bones are much the best, and you will be surprised at the quantity that they will eat.

In regard to feeding troughs, I have tried all that I ever saw that I thought was of any use and I have never found anything that is superior to a shallow tin pan for small chicks, they are easily kept clean and are the best, any kind of a trough that will let the head of a small chick through will let the body of all the smallest ones in and it is worse than useless.

As soon as the chick comes to be of any size you can use a trough with a row of round pins on each side one inch apart cut level on top so that you can cover it with a light board, this is a good trough for any kind of fowls old or young.

The most difficult part of the whole business is to water the chicks in such a manner that they will not get wet, no matter what kind of a vessel is used it must be large enough so that the chicks can all drink at once, else they will fight and all be wet before any of them can drink. The best drinking vessel is made in the following manner, they can be made of any desired size, I will give the proper size to make one large enough to water twenty-five chicks all at once, so that they will not fight and get wet.

Make a tin pan nine inches in diameter, with a rim turned at right angles and three-fourths of an inch deep, then make a second one-eight inches in diameter and two inches deep, solder three stops on one or the other so as to hold the eight inch pan at an equal distance from the nine inch one all round, then make three holes one-fourth of an inch in diameter in the eight inch pan, make them so that the upper edge of the holes will be just a half inch from the bottom of the pan.

Now, to fill them just invert the eight inch pan in a bucket of water, place the nine inch one over it and turn over with a quick motion, the water will always stand a half inch deep in the outside or nine inch pan but no deeper, so that the chicks can drink without getting wet, they are so easy to keep clean and are cheap, any tinner will make them for twenty cents each, I know of nothing that is equal to them, freezing will not burst them unless clear full then it will.

Eternal vigilance is the price of Liberty. Eternal vigilance is the price of chicks. The man or woman who is as regular as a clock in the feeding of and the care of the chicks will succeed, those who are not will fail, chicks must be fed early in the morning, and they must be fed in the evening before they want to go to sleep, they follow the old maxim, early to bed and early to rise, they should be fed as soon as they are up in the morning, a chick that is in good health and is fed properly will always have an empty crop in the morning, and it should always be full at night, if it is not it is either because the chick did not get what it wanted or else it is not in good health.

Always go round at night just about the time that the chicks are going to bed and see that they don't pile up, that is the time that they are the most apt to do it, just when they are going to roost take your hand and gently spread them out so that they will be sure to all get plenty of air, when they are a week old is the time that you will need to use the most care, before that

BROWN LEGHORNS.

they are not strong enough to hurt each other much, and after they get a few days over a week old they are not so apt to crowd.

If they will pile up in spite of you, why then the only remedy is to have the brooder so arranged that they cannot pile up, this can be done quite easily, just have a tank of water above them and raise them so close to it that they will have no room to pile up, but as I said before, if they are kept perfectly comfortable they will not do this, the only time that there is any danger is just when they want to go to sleep at first going to bed at night, you know that it don't feel very pleasant to get in a cold bed at night, and if the brooder is cold when the chicks go in it they will be sure to pile up and crowd each other, but if it is warm and comfortable they will not.

I wish to repeat, that if they are not kept warm and dry all else will fail, also, they must be kept clean, one of the worst things about them is that when confined they will eat the droppings of each other unless they are kept clean.

Chicks will never do any good if they are allowed to sweat in the brooder, the only way to prevent it is to give plenty of ventilation of warm dry air.

PROPER TEMPERATURE OF INCUBATORS.

The very best point is one hundred and three degrees, one hundred and four degrees, or one hundred and two degrees, either one will do, one hundred and five degrees will not kill the chicks but is too warm, it will not do to buy thermometers wherever you happen to come across them, they will be wrong nine times out of ten, you must either get your thermometer of some one that keeps them for that purpose or else you will have to experiment until you get the proper heat, those that are sold as a general thing are not correct unless a high price is paid for them.

The tubes should be seasoned for at least two years before they are marked, the ordinary weather thermometers are usually made with the marks so close together that it requires the greatest care in reading them or else you can vary two or three degrees and not know it.

Thermometers for hatching eggs should have the degree marks one-eighth of an inch or near that apart, so that the slightest change can be noted.

When eggs are first put in the incubator it will take much more lamp power to keep up the heat than it will after the germs begin to develop and the eggs to give off animal heat, they do not give off much until the end of the second week, the only proper way to get the exact heat of the eggs is to lay the thermometers on the eggs, and you must be very sure that it is on fertile eggs, eggs that are unfertile will not be near as warm as the fertile ones even if they are laying side by side.

When you first put the eggs in the incubator you cannot tell which is the fertile eggs and of course it will make no difference at that time, for all will be of the same temperature, but as soon as the germs begin to develop you must test them and find out which are the fertile ones, and then keep the thermometers on them, keep it at or near to one hundred and three degrees all the time from first to last, one hundred and four degrees will do as well, but you are just one degree nearer to the danger point and will not hatch one more chick at one hundred and four degrees than you will at one hundred and three degrees.

Don't ever lay fresh eggs in among those that are well advanced, it will destroy the evenness of the heat, if you wish to put eggs in at different times keep each batch by itself, and don't let the fresh eggs touch those that are under way.

The best results will be obtained by filling the machine full all at one time, and it will be much less trouble to care for the chicks, but it often will happen that a person wants to put in a few at a time, and they should know the plan that will give the best results, it is as I said before, keep each lot separate from the others.

All of my large machines, from six hundred eggs and upwards are now so made that by the use of a system of stop cocks whereby the circulation of the water can be controlled at the will of the operator, each set of trays can be kept at a different temperature, so that one large machine can be run in just the same manner that two or more small ones could, that is, one set of trays can be full of eggs that are well advanced and another full of fresh eggs and each set kept at the same heat which would be impossible under ordinary circumstances. And when it is desired to do so the entire machine can be kept at the same heat, the great advantage of this arrangement will be at once appreciated by any one that has ever run an incubator.

TESTING OF THE EGGS.

This is a subject of much importance, and it is one that can only be thoroughly learned by experience, although a little intelligent instruction will learn a novice more in one hatch than he would learn in two years by experience as a general thing.

First, all eggs must be divided into two classes, fertile and unfertile, the fertile eggs must be divided again, those that will hatch and those that will not, those that will hatch must be divided into three classes, viz:, those that are vigorous, those that are weak, and those that are very weak.

The unfertile ones should all be classed as one kind, but there will be an almost infinite variety of forms, these will be described presently.

The fertile ones (that is each of the three separate classes), will all look as much alike as two peas, or as so many grains of corn; of the vigorous, all will hatch, barring accidents; of the weak, most of them will hatch under good conditions; of the very weak, they can all be considered doubtful, some will hatch and some will not, these differences will not be distinctly marked until the sixth day, and to an operator that has but little experience will not be plain even then without instructions.

First, the vigorous, these on the third day of incubation will show a small ant shaped clot with the small, bright red blood vessels radiating in all directions, a ten cent piece will usually cover the whole thing at the age of three days, but I have seen them so large that a half dollar would hardly do it, with clear shells, a good light, and a good egg tester, the heart can be seen beating quite plainly on the third day.

I wish to remark here, that the best egg tester that can be made is in the reach and means of any one who will go to a little trouble to make it, all that is necessary is to make a room perfectly dark, or even darken a single window both inside and outside, it must be one that will get the sunshine, then have a board with a hole through it and a piece of soft gum or soft leather with an oblong hole, and made so that the eggs will not go through it when held tight against it to exclude the light.

No light must come in the room except what comes through this hole, and the sun must shine on the hole in order to get the best results, although fair work can be done on a cloudy day, for testing large numbers it is the only thing that I ever tried that would not tire my eyes, the main thing is to have the room perfectly dark.

Well, to resume, the vigorous eggs on the fifth day will look more like a large spider than anything else. although it is impossible to see one in a hundred of the blood vessels with the naked eye, still there will be enough of them seen to make it look like a large spider sitting on a small web, on the fourth day a black dot will be seen on the small end of the clot, on the sixth day this will be much increased in size, on the eighth day it will appear to be the largest part of the chick, this is the eyes.

On the fifth day the chick begins to move, it will only occasionally do this, on the sixth day it will move quite often, on the seventh day it will be in motion constantly, a waving pulsating movement, backwards and forwards, up and down in a circle apparently, never stopping more than a few seconds at a time, this motion I think is involuntary, and is no more under the control of the chick than is the beating of the heart, but I do not state this as a fact for I know of no means of proving it to my own satisfaction, this motion is kept up constantly until about the twelfth day, when it gradually begins to cease until on the fifteenth day the motion is entirely under the control of the will of the chick and it will only move when it wants to.

The circulatory system of an egg is one of the most beautiful objects in nature, thousands of minute air cells line the inside of the shell, these take the place of the lungs in purifying the blood, the blood is forced by the action of the heart into the circulation, does the perfect work of nature by contributing to the growth of the embryo, it is carried to the air cells, becomes purified and is then carried back to the heart, again to be sent on the same round of labor, this goes on constantly and evenly (that is, it will if the heat is even), and as all the air that the blood receives comes through the pores of the shell, it shows how important to the successful hatching of healthy chicks, it is that the eggs should be perfectly clean.

As soon as the blood vessels become well developed, an experienced eye will at once detect the difference between the main veins and arteries. The veins will be of a deep blackish red and the arteries will be a bright scarlet, the difference will not be so marked as the difference in venous and arterial blood that is taken from man or an animal, but it is quite plainly marked to an experienced eye.

Of the weak eggs the only apparent difference between them and the vigorous is that they will be very much smaller in size at a given age, and the blood vessels will not look so red or so plentiful as in the others, but still the most of them will hatch under very favorable conditions, but the chicks usually will be later coming out and when they are out they will not grow as fast as the others.

Fortunately if the eggs are from good, vigorous, healthy hens the proportion of weakly chicks will be very small.

Of the very weak, but few of these will hatch at all, the most of the very weak eggs will be loose, by that I mean that the blood vessels are not fast to the shell, this can be told by turning the egg over and over, and the embryo will be seen to float freely all through the egg, should it not be fast to the shell by the sixth day at the farthest, the chick will not hatch, it will be for want of air if for no other reason.

Most of them will be dead by the eighth day, but all that survive that period stand a chance of getting out if the circumstances are all favorable, if not they are sure to die.

Last, comes those that don't ever hatch at all, although fertile, these in poultry parlance are called addled, they are what make the rotten eggs, an egg that is unfertile will keep perfectly sweet in the incubator during the whole period of incubation, and if it is not shook or disturbed will keep longer even than that.

The addled eggs have such an infinite variety of forms that it would be out of the question to describe them all, they also come from various causes, mostly from imperfect germs, many are made by the hens sitting on the eggs long enough to just start the germ growing, the egg is then taken away and of course the growth stops at once, then, if the egg is put in the incubator in a day or two no harm will be done usually, but if it is kept any length of time the germ either dies outright or else when it is put in the incubator it will degenerate into some kind of a small monstrosity, live a few days and then die.

Some of them will appear as a small round ring, others as a large ring the size of a quarter to the half of the egg, others will appear to be a zig-zag streak, and will run in all sorts of shapes. Looking for all the world as if they had been struck with lightning, or rather like a zig-zag streak of lightning, others will appear to have started off all right, and then a small vein will have broken and the embryo has bled to death, some will show nothing but a dirty looking spot, others will just be slightly cloudy and no signs of red at all, some will show a long, straight streak, others a short one, again we will see one with a reddish band from one-eighth to one-fourth of an inch wide, running clear round the shell, this kind is not often seen, in my experience, not once in a thousand eggs, but it is not necessary to name any more varieties, when one is seen that is radically different from the spider-like form, it may as well be thrown away, for even if it is not dead it will either die in the shell or else it will hatch a cripple.

I would advise all operators to select out all but the perfect and vigorous embryos and throw them away for even if you get them hatched out they will not on the average pay for the raising.

It is much cheaper to throw a few eggs away than to bother with a few weakly chicks that will never pay for tho raising.

Of course it will not do for one that has no experience to do this for they would be apt to throw those away that might be good, but a years work at the business will tell any one what they want to do.

The eggs of the Duck, Goose, Turkey and Guinea, will be very much slower about starting off than hen eggs, and must not be classed with them at all.

They will hardly show that they are started at all on the third day, even to an experienced eye, and a novice would not be able to see anything, without a very close observation, nothing but a small cloudy spot will show, and this would be overlooked by many.

There will mostly be a few chicks that will die when full grown even under the most favorable condition, this will occur from a variety of causes, some of them will stick their beak through the skin of the yolk and that is sure death, others will choke either because the shell was too thick to allow it to dry up sufficiently to supply the chick with air when it needed it, or because it has the head in the wrong part of the shell.

Some will die from sheer overwork and exhaustion, trying to get out, especially if the shell is very thick and hard, they are very easily overheated at this time as the exertion of the chick in trying to get out will heat it more or less owing to how hard it works.

This is the time that the most care is required to see that the heat does not go too high, keep the thermometer right in among the chicks that are coming out and see that the heat does not go higher than one hundred and three degrees. The best heat for chicks that are coming out lively is one hundred and two degrees, and it is very important that the heat should be just right, it is just as bad for it to be too low as too high, if the heat is too low the chick will not properly absorb the yolk.

The majority of chicks absorb the yolk into the abdomen just as they are coming out, that is, they do this when they come out on time, if they are a day or two late coming out they may have the yolk absorbed before they chip the shell.

But if they come out on time, they first chip the shell and then take a rest of from two to ten hours, they then begin to turn around, and as they turn they throw back the head and each time they throw back the head they will break the shell a little farther until they finally break it so near around that by stretching out they are able to burst it open and come out. Well, if the heat is too low when this is going on it seems to operate against the absorption of the yolk so that the chick will come out with the yolk more or less out and it is usually useless to try to save it, while, if the heat is too high, it, with the exertion of the chick, will often overcome it so that it will die and nothing can be seen to be the matter with it, only too much heat

Now here is two rules to go by and either one will give you good results.

First, keep the heat at one hundred and three degrees all the time from the first day the eggs are put in the incubator until the chicks begin to chip the shells, then drop to one hundred and two degrees until they are all out.

Second, set the regulator so that the heat will run from one hundred and two degrees to one hundred and four degrees and back again, etc., from first to last, and as soon as they begin to chip set it so as to run steady at one hundred and two degrees.

I have no choice between the two methods as far as the number of chicks that will be hatched is concerned, but if there is any difference in the chicks it will be in favor of the latter method. Mind, I don't say that there is any difference, but I believe that the latter method allows more oxygen to reach the blood of the chick than the first, but this is a point that each one can decide to their own satisfaction, one way will hatch as many chicks as the other.

But which ever plan you may adopt, be sure above all things to not overheat them, nothing is more discouraging to an operator to hatch out a fine looking lot of chicks and have them all die and not know what is the matter, if they are properly hatched they will be the strongest kind of chicks, but if not, they may appear to be all right and still all die. If they have the proper heat and plenty of fresh air all the time and then come out on time they will be all right, and will, if properly fed, out grow nine-tenths of those that are hatched by hens. There is not more than one hen in ten that can keep the heat just right all the time, unless the weather is favorable.

THE
BEST INCUBATORS WE KNOW OF.

THE EUREKA INCUBATOR,

MANUFACTURED BY MR. J. L. CAMPBELL, WEST ELIZABETH, PA.

The cut represents an Incubator with a capacity of three hundred eggs. It is made as follows:

The case is made of wood. It is double, with an air space between the inner and outer case. This serves a double purpose, it makes it possible for the machine to withstand a large variation of temperature, and it also protects the outside case from the action of the heat and moisture from within.

The heat is distributed to the eggs by means of a double copper tank, the lower tank is made to cover the entire bottom of the machine and it is connected to the top tank by four or more brass pipes, owing to size of machine. A false bottom is placed inside over the lower tank, and between this and the tank is a hot air space where all the fresh air that goes inside the machine is heated, on the top of this false bottom is placed a pan of water which furnishes all the moisture that is required to hatch the eggs.

The water in the lower tank is heated by the lamp and the hot water rises to the top tank by means of the connecting pipes. The eggs are placed under the top tank and the greater part of the heat that hatches them is radiated from the upper tank, just enough is allowed to come from the lower one to overcome the bad results of having it entirely cold below the eggs.

The heat is regulated perfectly by a most simple and perfect arrangement that is entirely different from any other one that is in use at the present time.

Nothing is used that will ever wear out or need renewing. There is a thermostat placed inside just over the eggs and controls the exact heat of the eggs, a lever is connected to this thermostat and comes out on top of the machine as shown in the cut, this lever is moved one way by the heat and the other way by the cold, it is so sensitive to a change either way that the amount of variation required to move it cannot be measured on a common thermometer, it requires about the one-tenth of one degree to move the lever sufficient to make it shut off the lamp and open a ventilator. This is done as follows:

There is a small clock-work movement that is connected to the lamp, and to a valve or ventilator that is placed in the top of the machine, there is a lever on this clock that has a long and a short end, one end always rests on the top of the thermostatic lever. When the short end of it rests on the thermostatic lever the ventilator is closed and the lamp is burning full, as soon as the heat rises to the point at which it is set the thermostatic lever moves back a little and allows the short end of the clock lever to slip off, in doing this the clock lever makes a half revolution and the long end catches and at the same time opens the ventilator and turns down the lamp.

This can be set so as to hold the heat exactly at one point, or it can be set so as to allow the heat to run up and down from one degree, to two degrees, or even three degrees, but not over three.

Another special feature of this machine is the clock for turning the eggs, all that has to be done is to wind the clock twice a week and the eggs are all carefully turned four times each twenty-four hours, this is a very important feature of this machine, as the eggs are turned no matter whether the operator is present or not.

Underneath the machine is a nursery for the young chicks, they are taken out of the incubator as soon as hatched and placed in the nursery, this nursery is heated by the same lamp that heats the eggs, there is a hot air chamber in the centre of the nursery, the fresh air is conducted into this from the outside and becomes heated and is distributed over the top of the chicks.

The entire machine is very perfect and at the same time it is not at all complicated, any person of ordinary intelligence can comprehend it at once, it has the fewest parts of any self regulating machine that is made.

The machine does the work to perfection. Many who have bought them claim to have hatched every egg, but the maker considers that eighty-five per cent. is a good average, taking eggs as they come, good eggs that are well taken care of in storing will average that many all the time, at the same time it is a very common thing to hatch ninety-five per cent.

The machine is made in all sizes from sixty eggs to five thousand eggs, the large machines are much the cheapest in proportion to the number of eggs that they will hold, the labor of making one that will hold sixty eggs is just the same as to make one that will hold three hundred eggs. The prices for the complete machine is as follows:

60 eggs, $57; 100 eggs, $62.50; 200 eggs, $69; 300 eggs, $75; 500 eggs, $116; 600 eggs, $126; 1,000 eggs, $180; 1,500 eggs, $222; 2,000 eggs, $264; 2,500 eggs, $315; 3,000 eggs, $346; 4,000 eggs' $438; 5,000 eggs, $520.

INCUBATORS. 175

THE EUREKA BROODER.

The cut represents a brooder for two hundred chicks, the size of it is three feet wide and six feet long, twenty inches high at the comb of roof, the sides are made of two wide boards cut to the proper shape and the floor is nailed on cross ways out of flooring.

The heating apparatus consists of a double sheet-iron lamp chamber, a pair of hot water tanks and a lamp.

The lamp chamber serves the double purpose of heating the water in the tanks, and it also makes a hot air heater, the air enters at the bottom of the heater on the outside of the brooder and is heated and then rises to the top and comes out inside over the top of the chicks, the brooder can be used as a hot air brooder and not use the tanks if it is so desired, but it is best to use the tanks as a general thing.

The tanks rest on the top of the lamp box or chamber and are heated through the iron top, the under side of the tanks are covered with flannel, the lower part of the brooder is divided into four rooms, movable platforms are placed under the tanks so that the small chicks can get their backs up against the warm tanks.

The large room above the tanks answers the purpose of a nursery for small chicks.

In using the brooder the chicks must be kept warm enough so that they will not pile up on each other. Also, when it is so desired, the platforms can be raised up so close to the tanks that the chicks have not room to pile on each other, but it is always best to keep them warm and comfortable and then they will sit around and not crowd at all.

More failures to raise the chicks that are hatched in incubators have resulted from keeping too many in one lot, than all the other causes combined, fifty chicks in one lot is as many as should be kept together.

Poultry raisers are beginning to find this out, and those who have adopted the plan of keeping the chicks in lots of not over fifty chicks have been successful.

Nothing is of more importance than plenty of pure air without draughts. The Eureka Brooder uses the correct system of ventilation for this purpose.

All know that hot air rises, also, that foul air sinks, when the brooder is in operation the fresh air in going through the heater has considerable pressure, and coming in at the top it gradually mixes with the inside air so as to not cause a draught and at the same time it forces the foul air out at the floor of the brooder at the openings that are provided for that purpose.

This brooder can be used out in the open air and will keep the chicks dry and warm no matter how severe the storm, but all brooders should be used in a shed if possible.

The brooder shown in the cut is arranged so that four separate yards can be made to connect with it, and each one will have its own door to go in and out, each room is large enough to accommodate fifty chicks until they are large enough to do without artificial heat. Then the nursery can be made to accommodate one hundred chicks more until they are from a week to ten days old.

The nursery has bottom heat only, and while this is good for young chicks for the reason that there is no place for them to go out in the cold until they learn to go in under the tanks, still I never could successfully raise them to any size with bottom heat alone for the reason that it would make them weak in the legs almost invariably. Price list of Brooders are as follows :

25 chicks, $6 ; 50 chicks, $8 ; 100 chicks, $12 ; 200 chicks, $16 ; 500 chicks. $20 ; 1,000 chicks, $35.

THE KEYSTONE INCUBATOR AND BROODER.

MANUFACTURED BY THE KEYSTONE INCUBATOR COMPANY,

PHILADELPHIA, PA.

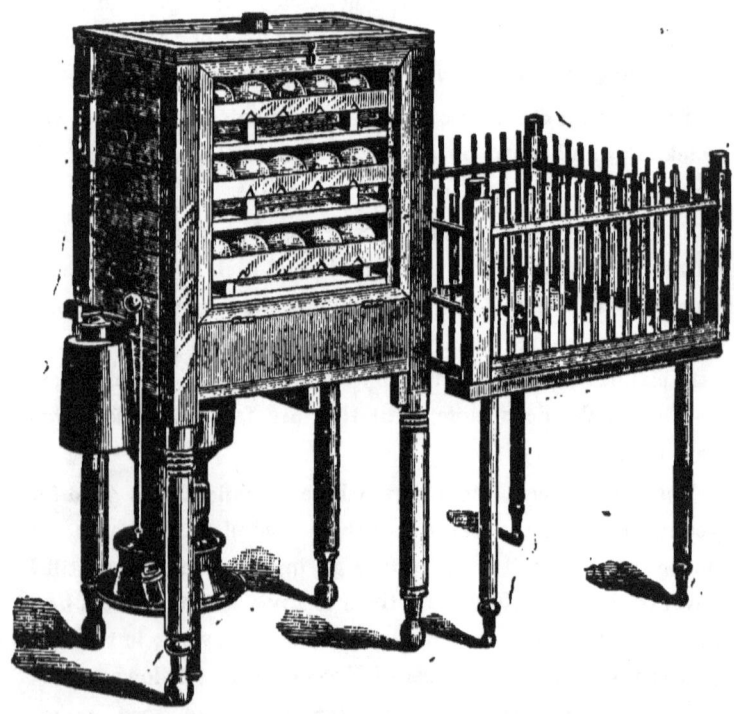

The following description gives a full and clear analysis of the construction of this machine, showing in detail its various parts and method of operation.

Heat is applied by means of water, which is heated by a small oil stove on the outside, and kept in constant circulation through tubes running upward in each of the four corners of the machine, and thence through a hollow framework, which is made of the best sheet copper, and serves as shelves for the egg frames. By this means the water is kept uniform in all parts of the machine, which is a matter of great importance in securing successful results, and is a distinctively original feature of this machine.

The hollow framework before mentioned is neatly fitted into a wooden case, with a door on one side, to facilitate the handling of the egg frames, and open at the bottom, which is flush with the lower hot water shelf.

A space of three inches, directly beneath the lower hot water shelf forms the brooder, which is open on one side for the attachment of the yard, as shown in the cut, which allows the chicks to run about at will.

The egg-frames are three in number, and are made to slide between the hot water shelves.

Beneath the egg-frames are the moisture-pans, the bottoms of which are lined with asbestos, which, by pouring on water, will take up and hold enough to supply the eggs with moisture for twenty-four hours.

The heat regulating apparatus is a thermostatic arrangement. The least variation in heat—so little that the thermometer does not register—will cause the regulator to act. When the temperature is too high it shuts off part of the flame, thus saving oil. The regulator can be seen to open and close frequently in a very short time, the mercury in the thermometer all the while remaining stationery.

This machine obtained first premium at the Pennsylvania State Fair, held in Philadelphia in September 1884.

Prices are as follows: 100 eggs, $50; 200 eggs, $60; 300 eggs, $75. Yard for brooder $5 extra.

SAVIDGE HYDRO-INCUBATOR.

MANUFACTURED BY C. W. SAVIDGE, 2524 HUNTINGTON STREET,

PHILADELPHIA, PA.

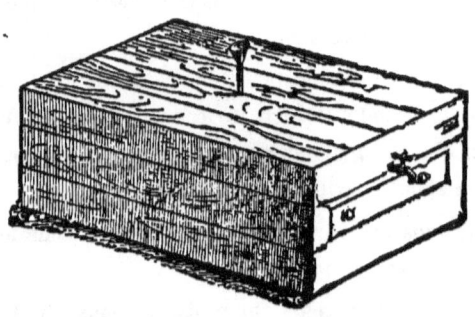

The above shows what the incubator looks like when finished. It is a tank enclosed with four inches of sawdust. The sawdust, which is called the packing, is *under*, *all around*, and on the *top*, even in front of the egg-draw, so that the heat is completely retained and enclosed. On the top is a tin tube, into which the water is poured into the tank in the inside, and in front, just above the egg-drawer, is a spigot by which the cold water can be drawn off before adding that which is hot.

The sizes and dimensions need not be given, for the reason that some prefer small sizes while others are partial to those that are larger. A tank fifteen inches wide and twenty-five inches long will answer for an incubator holding seventy-five eggs. The tank should be twelve inches deep. Any depth may be given it that may be desired, for the *greater the volume of water, the easier and longer the retention of heat.*

DIRECTIONS FOR RUNNING THE SAVIDGE INCUBATOR.

Place the incubator in a room where the temperature does not fall below fifty degrees Fahrenheit, if possible. This is not absolutely necessary, but it will be found more economical, and then fill the tank with boiling water; it must remain untouched for twenty-four hours, as it requires time during which to heat completely through; as it will heat slowly it will also cool slowly. Let it cool down to one hundred and fifteen degrees, and then put in the eggs, or, what is better, run it without eggs for a day or two in order to learn it, and notice its variation. When the eggs are put in, the drawer will cool down some. All that is required then, is to add about a bucket of water once a day in summer, and in the winter once or twice a day, in the morning and at night, but be careful about endeavoring to get up heat suddenly, as the heat does not rise for five hours after the additional bucket of water is added. The tank radiates the heat down on the eggs, there being nothing between the iron bottom of the tank and the eggs, for the wood over and around the tank does not extend across the *bottom* of the tank. The cool air comes from below in the ventilator pipes, passing through the muslin bottom of the egg drawer to the eggs. Lay the eggs in, the same as in a nest, promiscuously. Keep the heat inside the egg drawer as near one hundred and three degrees as possible. Avoid opening the egg drawer frequently, as it allows too much escape of heat. *Be sure your thermometer records correctly*, as half the failures are due to incorrect thermometers, and not one in twenty is correct. Place the bulb of the thermometer even with the top of the eggs, that is, when the thermometer is laying down in the drawer, the upper end should be slightly raised so as to allow the mercury to rise, but the bulb and eggs should be of the same heat, as the figures record the heat in the bulb and not in the tube. Keep a pie pan filled with water in the ventilator for moisture, and keep two or three moist sponges in the egg drawer, displacing a few eggs for the purpose. Turn the eggs half way round twice a day at regular intervals—eight o'clock in the morning and six o'clock at night. Let the eggs cool down while turning them; it will do no hurt; but do not let them cool lower than seventy degrees. No sprinkling is required if the sponges are kept moist. If the heat gets up to one hundred and ten degrees, or as low as sixty degrees for a little while, it is not necessarily fatal. Too much heat is more prevalent than too little. A week's practice in operating the incubator will surprise one how simple the work is. Heat the water in one or two boilers, as a large quantity will be required when *filling the first time*, and pour it in through the tube on top of the incubator, boiling hot, using a funnel in the tube for the purpose. Just at the time of hatching out do not be tempted to frequently open the drawer. Cold drafts are fatal. Patience must be exercised. When turning eggs in the morning let them cool well. The best results have been obtained when the heat was kept at one hundred and five degrees the first week, one hundred and four degrees the second week, and one hundred and two degrees the third week.

Prices of the Savidge Incubator, 100 eggs, $21. No charge for boxing. 50 eggs, $18. Brooder, $10.

THE "PERFECT" HATCHER.

MANUFACTURED BY THE PERFECT HATCHER CO., ELMIRA, N. Y.

This is a self regulating incubator; the water tank is above the eggs, and it is warmed by a heater exterior to the machine, the nursery is placed underneath the egg trays, and the method of heating is by gas or kerosene oil.

The heat is regulated by means of a thermostatic bar, which, by moving a lever, connects an electrical battery with a clockwork, which, when the heat reaches the highest point, opens a ventilator and reduces the flame of the lamp, and when the lowest permitted temperature is reached, closes the draft and turns up the flame.

Thus the heat constantly, but slowly vacillates between the two extremes as fixed.

A peculiarity of this machine is the double layer of eggs. The upper tray is placed below, after being set ten days, and then the upper one is filled.

The battery connected with this machine is warranted to work a year without renewal; the clock needs to be wound but once in three days.

The "Perfect" Hatcher is very highly commended by those who use it and is said to be very certain and satisfactory in its results.

THE "NEW CENTENNIAL" INCUBATOR.

MANUFACTURED BY A. M. HALSTEAD, RYE, N. Y.

This is a self regulating machine, very similar in its working to the "Perfect Hatcher." Mr. Halstead, the manufacturer of it, is one of the oldest and most successful incubator men in the country. It is well made, easily managed and gives universal satisfaction.

Of course, even with the successful machines described in this book, a person can easily fail to make a success. Perhaps it will be well to name a few things necessary to success: A steady, settled purpose to *make* the thing succeed. Intelligent study of the instructions supplied with the machine, so as to enable you to thoroughly understand it, just as an engineer understands his locomotive. Careful attention to cleanliness in the lamps, wicks trimmed evenly, etc. Patience for results. This will enable you to put twenty eggs into a one thousand egg machine for a start, and practically learn "how to do it." In a word use common sense, as much of it as you can command, and you are sure to come out successfully.

HOW TO KILL AND DRESS FOWLS.

Never kill your fowls until they have fasted twenty-four hours. No man ever made any money by selling his fowls with

BLACK-BREASTED RED GAMES.

their crops stuffed to make them weigh. The petty fraud is too apparent. To kill and dress, tie their legs together, hang the fowl up, open the beak and pass a sharp pointed, narrow bladed knife into the mouth and up into the roof, dividing the membrane. Death will be instant. Immediately cut the throat by dividing the arteries of the neck and the bird will bleed thoroughly.

We never scald; the nicest way is to pick the fowl dry and while yet warm. A little care will prevent tearing the flesh, and the bird will bring enough extra in the market to make it pay. Most persons, however, will prefer to scald, and for home consumption, or the village market this will do.

Have the water just scalding hot—*not boiling*—one hundred and ninety degrees is just right. Immerse the fowl, holding it by the legs, taking it out and in, until the feathers slip easily. Persons become very expert at this, the feathers coming away by brushing them with the hand, apparently. At all events, they must be picked clean. Hang turkeys and chickens by the feet, and ducks and geese by the head, to cool. It should be unnecessary to say that under *no* circumstance whatever, should ducks and geese be scalded; they must invariably be picked dry. Take off the heads of chickens as soon as picked, tie the skin neatly over the stump, draw out the insides carefully, and hang up to cool. Never sell fowls undrawn. They will bring enough more drawn and nicely packed, with the heart, gizzard and liver placed inside each fowl, to pay for the trouble. Let them get thoroughly cool—as cold as possible—but never, under any circumstances, frozen. There is always money in properly prepared poultry; the money is lost in half fitting them for market, the fowls often being forwarded in a most disgusting state. There is money in the production of eggs; there is money in raising poultry for the market. The money is lost in a foolish attempt occasionally made to make the buyer pay for a crop full of musty corn. at the price of first-class meat. It is that class of men, however, who are too smart ever to make money at anything.

PACKING FOR MARKET.

The poultry, having been killed as directed, carefully picked, the heads cut off, and the skin drawn over the stump and neatly tied—or if preferred, leave the head on, the fowl will not bring less for it—and the birds chilled down to as near the freezing point as possible, provide clean boxes and place a layer of clean hay or straw quite free from dust, in the bottom. Pick up a fowl, bend the head under and to one side of the breast bone, and lay it flat on its breast, back up, the legs extending straight out behind. The first fowl to be laid in the left hand corner. So placed, lay a row across the box to the right, and pack close row by row, until only one row is left, then reverse the heads, laying them next the other end of the box, the feet under the previous row of heads. If there is space left between the two last rows, put in what birds will fit sideways. If not, pack in straw at the sides and between the birds, so they cannot move. Pack straw enough over one layer of fowls, so that the others cannot touch, and so proceed until the box is full. Fill the box full. There must never be any shaking, or else the birds will become bruised, and loss will ensue. Many packers of extra poultry place paper over and under each layer before filling in the straw. There is no doubt but that it pays. Nail the box tight; mark the initials of the packer, the number of fowls and variety, and mark plainly the full name of the person or firm to whom it is consigned, with street and number on the box, Thus the receiver will know at a glance what the box contains, and does not have to unpack to find out. These directions, if carefully carried out, might save a person many times the cost of this book, every year.

KEEPING EGGS FRESH.

The place selected for keeping eggs should be cool in summer, but not cold in winter, that is, it should be kept at a temperature of from forty-five to sixty degrees all the year round. If too cold, the eggs will freeze and crack, if too warm, they will commence to decay and get stale sooner than they otherwise would. Shelves should be fitted up with holes bored in them, sufficiently large to keep the eggs upstanding, but, of course, not large enough to allow them to pass through. These shelves will be very inexpensive and will serve a lifetime. The eggs should be placed in these holes broad-end *downwards*, and tests have proved that they will keep fresh in this position, very much longer than with the broad-end upwards.

INDEX.

		Page
Age, to Breed,	Part 2,	42
" to Fatten,	" 2,	42
" to Judge,	" 2,	42
" to Kill,	" 2,	43
" to Show,	" 2,	43
Air Bubble or Cell,	" 2,	44
Alarm,	" 2,	44
Ale,	" 2,	44
Appetite,	" 2,	46
" Loss of	" 2,	109
April, Work for	" 2,	46
Artificial Incubation,	" 2,	47, 144
" " Advice to Beginners,	" 2,	145
Artificial Rearing,	" 2,	49
Asiatic Breeds,	" 2,	50
" Diseases,	" 2,	51
Asphalt Flooring,	" 2,	51
Atropine,	" 2,	51
August, Work for	" 2,	51
Awards,	" 2,	52
Bantams,	Part 1,	32
" Varieties	" 1,	32
Bad Feathering,	" 2,	13
Barn Door Fowls,	" 2,	54
Bedding for Chicks,	" 2,	54
Benzine,	" 2,	55
Best Breeds for Market,	" 1,	67
" " Confinement,	" 2,	71

Black Hamburgs,	Part 2,	55
Bone Dust,	" 2,	55
Brahma Fowls,	" 1,	3
Breast-Bone, Crooked	" 2,	56
Breeding Cocks,	" 2,	56
Breeding In-and-in	" 2,	56, 104
Broeds for Crossing,	" 2,	57
Broken Leg,	" 2,	22
" Bones,	" 2,	58
Bruises and Fractures,	" 2,	21
Buying Poultry and Eggs,	" 2,	59
Carbolic Acid,	Part 2,	61
Capons and Caponizing,	" 2,	61
Caponizing,	" 2,	141
Change of Place,	" 2,	63
Chicks, Ordinary Fattening of	" 2,	94
Chickens, General Treatment of	" 2,	64
" Precocious,	" 2,	119
Combs, Torn	" 2,	25
" Varieties of	" 2,	69
" Pea,	" 2,	117
" Wire Support to	" 2,	140
Cochins,	" 1,	8
" Varieties of	" 1,	9
Cocks,	" 2,	67
Cockerels,	" 2,	67
Cockerels and Pullets, Separation of	" 2,	129
Cold Weather,	" 2,	68
Concrete Flooring,	" 2,	70
Condition to keep Fowls in,	" 2,	71
Cooking for Poultry,	" 2,	72
Corns,	" 2,	23
Cramming,	" 2,	73
Crevecœurs,	" 1,	44
Daily Routine,	Part 2,	32
Defects, Choice between	" 2,	66
Dirt and Droppings,	" 2,	78
Dictionary of Poultry Terms,	" 1,	62
Diseases of Poultry,	" 2,	4
" " Apoplexy,	" 2,	13
" " Appetite, Loss of	" 2,	109
" " Asthma,	" 2,	12
" " Bumble Foot,	" 2,	24
" " Black Rot,	" 2,	55
" " Canker,	" 2,	60
" " Catarrh,	" 2,	63

Diseases of Poultry, Cholera,			Part 2,	7,	66
"	"	Colds,	" 2,		68
"	"	Cramp,	" 2,	20,	74
"	"	Cramped Sitting Hen,	" 2,		74
"	"	Crooked Breasts,	" 2,		75
"	"	Crop-bound,	" 2,	15,	75
"	"	Dropsy,	" 2,		17
"	"	Diarrhœa,	" 2,	8,	78
"	"	Eruptions,	" 2,		89
"	"	Elephantiasis,	" 2,		23
"	"	Gapes,	" 2,	10,	99
"	"	Giddiness,	" 2,		100
"	"	Gout,	" 2,		11
"	"	Hereditary,	" 2,		103
"	"	Hoarseness,	" 2,		103
"	"	Indigestion,	" 2,	9,	104
"	"	Inflammation of the Lungs,	" 2,		104
"	"	" " Mucous Membrane,	" 2,		10
"	"	Influenza,	" 2,		18
"	"	Liver,	" 2,		108
"	"	Leg-Weakness,	" 2,	10,	107
"	"	Loss of Feathers,	" 2,		96
"	"	Megrims,	" 2,		12
"	"	Mange,	" 2,		16
"	"	Moulting,	" 2,		14
"	"	Paralysis,	" 2,	19,	116
"	"	Pip,	" 2,	18,	118
"	"	Phthisis,	" 2,		20
"	"	Rheumatism,	" 2,	20,	123
"	"	Roup,	" 2,	6,	125
"	"	Rump-Gland,	" 2,		22
"	"	Torpid Gizzard,	" 2,		17
"	"	Ulcers,	" 2,		23
"	"	Vertigo,	" 2,		138
"	"	White Comb,	" 2,		24
"	"	Wheezing,	" 2,		140
Dominiques,			" 1,		31
Dorkings,			" 1,		16
Drowning,			" 2,		22
Drainage,			" 2,		79
Drink,			" 2,		79
Ducks,			" 1,		56
"	Aylesbury		" 2,		53
"	Breeding Pens for		" 2,		57
"	Buenos Ayres		" 2,		59
"	Carolina		" 2,		62
"	Pekin		" 2,		117
"	Rouen		" 2,		124

Dubbing,	Part 2,	88
Earth Deodoriser,	Part 2,	88
Ear-Lobes,	" 2,	81
Early Roosting,	" 2,	82
Economy,	" 2,	82
Eggs,	" 2,	83
" Addled	" 2,	41, 83
" Blood-stained	" 2,	83
" Broken in nest	" 2,	83
" Bound	" 2,	15, 84
" Chilled	" 2,	65, 84
" Dropping	" 2,	81
" Ducks, Color of	" 2,	84
" Eaters	" 2,	88
" Fertility of	" 2,	84
" How to collect	" 2,	35
" How to Preserve	" 2,	37
" How to Pack	" 2,	40
" Hens, Color of	" 2,	84
" Keeping fresh	" 2,	187
" Nest	" 2,	112
" Moistening	" 2,	86
" Packing for Hatching	" 2,	115
" Preserving for Sitting	" 2,	86
" Preserving for Winter use	" 2,	86
" Producers	" 1,	67
" Production increasing	" 2,	85
" Selecting for Sitting	" 2,	87
" Soft	" 2,	87
" Sex of	" 2,	129
" Testing	" 2,	136
" Test of fresh	" 2,	88
" Unfertile	" 2,	83, 137
" Yolkless	" 2,	88
Epsom Salts,	" 2,	89
Exhibition Birds, Care of	" 2,	61
" " Selection of	" 2,	128
" " Washing	" 2,	138
Exhibition, Treatment after	" 2,	136
Fattening,	" 1,	97
" A la Bresse	" 2,	94
Fancy Points,	" 2,	90
Farm Poultry,	" 2,	91
Farmyard Duck.	" 2,	93

Feathers,	Part 2,	95
" Hard	" 2,	102
" Loss of	" 2,	96
" Laced	" 2,	106
" Eating	" 2,	95
" Paint and Tar on	" 2,	25
" Worm eaten	" 2,	140
Feathering of the Legs and Feet	" 2,	96
Felt,	" 2,	96
Fencing wire,	" 2,	97
Feeding, Excessive	" 2,	89
First outlay,	" 1,	70
Fleas,	" 2,	97
Flesh,	" 2,	97
Floors,	" 2,	98
Food,	" 1,	76
" Allowance of	" 2,	44
" Animal	" 2,	45
" After Exhibition	" 2,	98
" " Kinds to procure	" 2,	46
" Barley	" 2,	58
" Bran	" 2,	56
" Buckwheat	" 2,	59
" Buttermilk	" 2,	59
" Brewers grain	" 2,	105
" Cabbage	" 2,	60
" Corn	" 2,	73
" During Moult	" 2,	98
" Green	" 2,	101
" Hemp	" 2,	103
" Indian corn	" 2,	104
" Insects	" 2,	105
" Oats	" 2,	114
" Oatmeal	" 2,	113
" Onions	" 2,	114
" Parsnips	" 2,	117
" Quality of	" 2,	123
" Rice	" 2,	123
" Sour milk	" 2,	128
Foods, List of	" 2,	108
Fountains,	" 2,	98
Fowls, Common	" 2,	70
" Bad habits of	" 2,	27
" General treatment of	" 2,	29
" Fattening	" 2,	93
" Handling of	" 2,	102
" How to kill and dress	" 2,	184
French Breeds,	" 1,	44
Fresh Blood,	" 2,	98

Fright,   Part 2,   99
Frost-bite,   " 2,   99

Game Fowls,   " 1,   26
Geese,   " 1,   54
 "   Fattening   " 2,   95
Grass,   " 2,   100
 "   Run   " 2,   100
Gravel,   " 2,   100
Ground, Cleanliness of   " 2,   101
Guinea Fowl,   " 1,   52
 "   "   " 2,   101

Hamburg Fowls,   " 1,   33
Hardiness,   " 2,   102
Harm to Crops,   " 2,   102
Hatching, Early   " 2,   85
Hay,   " 2,   102
Heat,   " 2,   103
Hens, Age to lay   " 2,   41
 "   Broody   " 2,   58
 "   Setting, Their management   " 2,   130
 "   To prevent setting   " 2,   119
Hospital,   " 2,   104
Houses and Yards,   " 1,   73
 "   Early opening of   " 2,   81
Houdans,   " 1,   46

Incubation,   " 1,   88
 "   Artificial   " 2, 47, 144
Incubators,   " 2,   144
 "   Average time to hatch   " 2,   151
 "   Best place to run   " 2,   147
 "   Care of chickens   " 2,   154
 "   Eureka   " 2,   171
 "   "   Brooder   " 2,   175
 "   Feeding   " 2,   156
 "   Keystone   " 2,   178
 "   New Centennial   " 2,   184
 "   Proper temperature of   " 2,   161
 "   Perfect Hatcher   " 2,   183
 "   Savidge   " 2,   180
 "   Testing the Eggs   " 2,   163
 "   The best we know of   " 2,   171
 "   Watering   " 2,   156

Journeys,   " 2,   105

INDEX. 195

Killing for Table,	- - - -	Part 2,	105
Lameness,	- - - - -	" 2,	106
Langshans,	- - - - -	" 1,	37
La Fleche Fowls,	- - - -	" 1,	45
Laying, To prevent	- - - -	" 2,	107
Laying Mixtures,	- - - -	" 2,	106
Leghorn Fowls,	- - - -	" 1,	41
Lime,	- - - - -	" 2,	107
Limit to Numbers,	- - - -	" 2,	107
Lobes,	- - - - -	" 2,	108
Malay Fowls,	- - - - -	Part 1,	25
Management of Chicks,	- - -	" 1,	91
Manure,	- - - - -	" 2,	109
March, Work for	- - - -	" 2,	109
Marking Chickens,	- - - -	" 2,	110
Mating,	- - - - -	" 2,	110
May, Work for	- - - -	" 2,	110
Meals,	- - - - -	" 2,	111
Meat Producers,	- - - -	" 2,	111
Milk for Chickens,	- - - -	" 2,	111
Moult, To hasten	- - - -	" 2,	111
Moulting,	- - - - -	" 2,	14
Nests for Laying,	- - - -	Part 2,	112
Night Accommodation for Chicks,	-	" 2,	113
Number of Ducks to Drake,	- - -	" 2,	113
Number of Hens to Cock,	- -	" 2,	113
Old Chicks,	- - - -	Part 2,	114
Old Fowls for Table,	- - -	" 2,	114
Orchards,	- - - - -	" 2,	115
Over Fattening,	- - - -	" 2,	115
Packing Fowls for Market,	- - -	Part 2,	186
Parasites,	- - - - -	" 2,	116
Pen for Breeding Stock,	- - -	" 2,	118
Perch,	- - - - -	" 2,	118
Points of Poultry,	- - - -	" 1,	61
Poultry Keeping,	- - - -	" 1,	65
" " Failures in	- - -	" 2,	90
Plucking,	- - - - -	" 2,	119
Plumage,	- - - - -	" 2,	119
Poland Fowls,	- - - - -	" 1,	20
Plymouth Rocks,	- - - -	" 1,	34
Prize Poultry,	- - - -	" 2,	121
" " Their Treatment,	- -	" 2,	120
Produce Hatched, Good average	- -	" 2,	121

Pullets, - - - - -	Part 2,	122
" Not Laying - - -	" 2,	122
Purity, To be preserved - - -	" 2,	122
Repletion, - - - - -	Part 2,	123
Roofs, - - - . - -	" 2,	124
Roosting, - - - - -	" 2,	124
Runs, - - - - - -	" 2,	126
" Aspect of - - - -	" 2,	51
Sand, - - - - - -	Part 2,	127
Sawdust, - - - - -	" 2,	128
Sex of Chicks, Foretelling - - -	" 2,	85
Sheds, - - - - -	" 2,	129
Shelf under perch, - - . -	" 2,	129
Shelter Hurdles, - - - -	" 2,	130
Slipped Wing, - - - - -	" 2,	133
Small Yards, Sorts for - - -	" 1,	68
Snow, - - - - - -	" 2,	133
Soft Eggs, - - - - -	" 2,	16
Spanish Fowls, - - - -	" 1,	10
Spring Chickens, - - - -	" 2,	134
Squirrel Tail, - - - - -	" 2,	134
Stains on Feathers, - - - -	" 2,	134
Strain, To Commence - - - -	" 2,	134
Straw, - - - - -	" 2,	135
Sulky Cocks, - - - - -	" 2,	135
Sultans, - - - - -	" 1,	23
Sunflower, - - - - -	" 2,	135
Tonics, - - - - -	Part 2,	135
Traveling, - - - - -	" 2,	136
Trussing, - - - - -	" 2,	136
Turkeys, - - - - -	" 1,	47
" Fattening - - -	" 1,	50
Varieties, - - - - -	Part 1,	3
Ventilation, - - - - -	" 2,	137
Vermin, - - - - -	" 2,	25
Want of Condition, - - -	Part 2,	138
Warmth, - - - - -	" 2,	138
Wasters, - - - - -	" 2,	140
Weakness after Sitting, - - -	" 2,	140
Wyandottes, - - - -	" 1,	39

CONFESSIONS

—OF—

JEAN JACQUES ROUSSEAU

A New Edition of this Famous Book—Translated from the French—with 13 Illustrations by Ed. Hedouin.

625 Pages, Long Primer Type.

Large 12mo, Red Cloth, Paper Label, $3.50

SENT PREPAID ON RECEIPT OF PRICE.

WILLIAM L. ALLISON,

216 WILLIAM ST.,   N. Y. CITY

An International Jury of twenty-five members, at the PARIS EXPOSITION, awarded a Diploma of Honor,

THE GRAND PRIX
to the
SMITH PREMIER TYPEWRITER.

NO HIGHER AWARD WAS POSSIBLE,

and in the language of the JURY'S REPORT, it was given *"FOR GENERAL SUPERIORITY OF CONSTRUCTION AND EFFICIENCY."*

THE SMITH PREMIER TYPEWRITER CO.,
337 BROADWAY, NEW YORK.

www.ingramcontent.com/pod-product-compliance
Lightning Source LLC
Chambersburg PA
CBHW030743230426
43667CB00007B/818